I SEE MATH
THE BASICS

Think Visually -
Discover Your
Math Brilliance!

Adding, Subtracting,
Multiplying, Dividing,
Fractions, Decimals, Percents

Debra Anne Ross LAWRENCE
David Allen LAWRENCE

GLACIERDOG PUBLISHING

"God created everything by number, weight and measure."
Sir Isaac Newton

GlacierDog Publishing

A Division of

GlacierDog Intergalactica

Anchorage, Alaska

~ I SEE MATH: THE BASICS ~

THINK VISUALLY - DISCOVER YOUR MATH BRILLIANCE

Notice to the Reader

The authors and publisher have done their best to present accurate, error-free information in this book. Nevertheless, some errors may have escaped our notice. If you believe you have found an error in this book, please let us know so we can amend future editions. You can contact us through our website: glacierdogpublishing.com.

Table of Contents for Print Book

Acknowledgements

Much thanks to Maggie Ross for carefully editing our manuscript and for offering numerous helpful comments. We are also very grateful to her for creating the charming and expressive duck drawings and for her generous endorsement.

We are truly thankful for Professor Channing Robertson's meticulous review of our manuscript and for his insight and ideas for making this a better book. We also deeply appreciate his kind endorsement. Professor Roberson has given us much sagacious advice over the years. We respect his amazing expertise and appreciate his enduring friendship.

Dedication

To all you who are discovering your math brilliance!
Yes. That means you!

Introducing
I SEE MATH: THE BASICS

I don't get math! It doesn't make sense to me. I'm not good at it! All those chicken scratches make me dizzy. Maybe I'm dyslexic...or ADD...or is it ADHD? Whatever...

The usual way math is taught doesn't work for everyone. Have you ever SEEN math taught visually?

No...can you do that? Could it really help me learn?

Yes! You can learn to SEE math. **I SEE MATH: THE BASICS** explains basic math step-by-step using pictures and diagrams. Jump in the math pond with me! You just may have undiscovered math brilliance!

In this book you will see adding: 8 + 6 + 3 = 17 ...

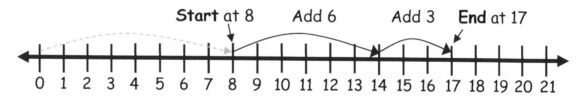

Start at 8   Add 6   Add 3   **End** at 17

0 1 2 3 4 5 6 7 8 9 10 11 12 13 14 15 16 17 18 19 20 21

...and subtracting, even when the second number is bigger: 2 – 4 = -2

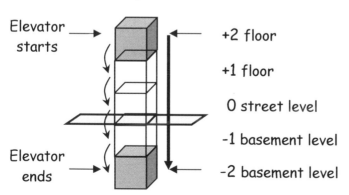

Elevator starts

Elevator ends

+2 floor

+1 floor

0 street level

-1 basement level

-2 basement level

Down -4 floors is +2 – 4 = -2

You'll see multiplication as adding the same number over and over...

12 Eggs multiplied by 3 = 12 + 12 + 12 = **36** Eggs

...and division as how many 4-egg omelets you can make with 12 eggs...

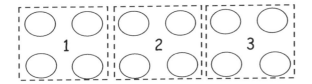

You can divide **12** eggs
into **3 groups of 4** to make 3
4-egg omelets. 12 ÷ 4 = 3

You will see a fraction as a part of a whole...

$$\frac{\text{Part}}{\text{Whole}} \qquad\qquad \frac{\text{1 Piece of the Pie}}{\text{Whole Pie}}$$

...and a percent as a fraction and as a decimal...

1% = 1 perCENT = 1 per100 = 1/100 = 1 hundredth = 0.01

hundredths
place

I SEE MATH: THE BASICS uses picture-based teaching. It will:

• Boost all students' understanding and grasp of math.

• Help parents tutor children and teens - great for home schools.

• Teach basic math to older learners who need a stronger math foundation for employment or further education.

• Show you: Numbers (natural, whole, integers, and negative); Number Lines; Base Ten; Addition; Subtraction; Multiplication; Division; Fractions; Decimals; Percents; and Converting Between Fractions, Decimals, and Percents.

Chapter 1
Numbers and the Number Line

So teach us to **number** *our days, that we may apply our hearts unto wisdom. Psalm 90:12*

1.1. Numbers: Natural, Whole, Integers, Negative, and Zero
1.2. The Base Ten System
1.3. Even and Odd Integers
1.4. Practice Problems

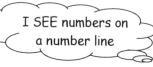

I SEE numbers on a number line

How many? How much? How long? How fast? How hot? Numbers are the best way to write down counts, amounts, and measurements. You can always picture a number by thinking about how far it is from zero on a number line. Every number represents a point on a number line. Every point on a number line can be expressed as a number.

The Number Line

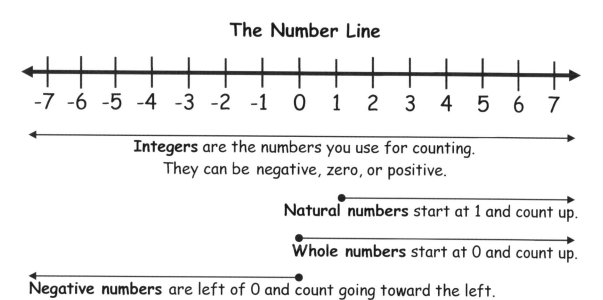

$$-7 \quad -6 \quad -5 \quad -4 \quad -3 \quad -2 \quad -1 \quad 0 \quad 1 \quad 2 \quad 3 \quad 4 \quad 5 \quad 6 \quad 7$$

Integers are the numbers you use for counting.
They can be negative, zero, or positive.

Natural numbers start at 1 and count up.

Whole numbers start at 0 and count up.

Negative numbers are left of 0 and count going toward the left.

1.1. Numbers: Natural, Whole, Integers, Negative, and Zero

Whole Numbers and Natural Numbers

The number line below shows
whole numbers and **natural numbers**

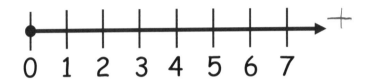

Natural numbers start at 1 and count up.

Natural numbers include **1, 2, 3, 4, 5, and up**.

Whole numbers start at 0 and count up.

Whole numbers include **0, 1, 2, 3, 4, 5, and up**.

Whole numbers include 0 and all natural numbers.

Whole numbers and natural numbers are the numbers you use when you count on your hands.

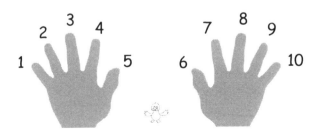

Integers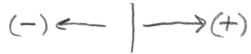

This number line shows **integers**

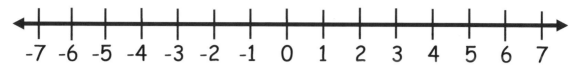

Integers include ... **-3, -2, -1, 0, 1, 2, 3,** ...

Integers can be **negative, zero,** or **positive**.

Zero is both an integer and a whole number.

Do you see why whole numbers and natural numbers are also integers?

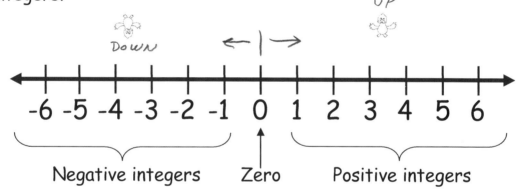

Negative integers   Zero   Positive integers

Negative integers start on the left side of 0 with -1 and count away from 0 to the left. Negative integers are opposites of the positive integers.

Positive integers start on the right side of 0 with +1 and count away from 0 to the right. Each positive integer has an opposite (negative) on the other side of zero.

Extra info: Fractions and decimals, such as 1/2 or 0.5, exist between all integers on the number line.

Zero

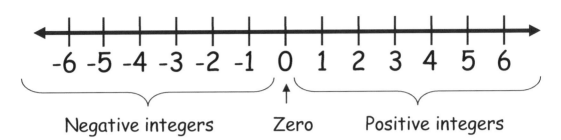

Negative integers   Zero   Positive integers

Zero means there is no quantity or value.

Zero is not positive or negative.

Zero is used as a place holder when there is no value.

For example, the number 101 is made up of

one hundred, zero tens, and one one:

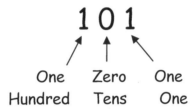

1 1 0 1
↑
THOUSAND

One   Zero   One
Hundred Tens   One

Extra information about zero: If you have a number and you...

...Add or subtract zero, the number won't change:

number + 0 = the number   or   **number – 0 = the number**

...Multiply by zero, the result is zero: **number x 0 = 0**

A number times zero is a number zero times, which is 0.

...Divide zero by the number, the result is zero: **0 ÷ number = 0**

Nothing divided by a number is still nothing.

...Divide the number by zero, the result is undefined:

number ÷ 0 = undefined

A number can't be divided by zero since zero has no value.

Consecutive Integers

Consecutive integers increase from the smallest to the largest (without any missing).

The hearts shown below increase in number and can be written:

{ 4, 5, 6, 7, 8 }

Other examples of consecutive integers are:

{0, 1, 2, 3, 4, 5}

{99, 100, 101, 102, 103, 104, 105, 106, 107, 108, 109, 110, 111, 112}

Consecutive integers include negative numbers and zero:

{ -8, -7, -6, -5, -4 }

Other examples of consecutive integers are:

{-2, -l, 0, 1, 2, 3, 4, 5}

{-20, -19, -18, -17, -16, -15, -14, -13, -12, -11, -10, -9, -8, -7}

Examples of lists of integers that are **not** consecutive:

{7, 9, 8, 10}   the 9 should come after the 8

{2, 5, 6, 9}   the 3, 4, 7, and 8 are missing

1.2. The "Base Ten" System

The **base ten** system is the system of numbers we usually use for counting and math. (It fits with our ten fingers.)

We can organize numbers using columns or "places" by tens.

The drawing below begins with the ones place on the right, and increases by *ten times* for each place to the left:

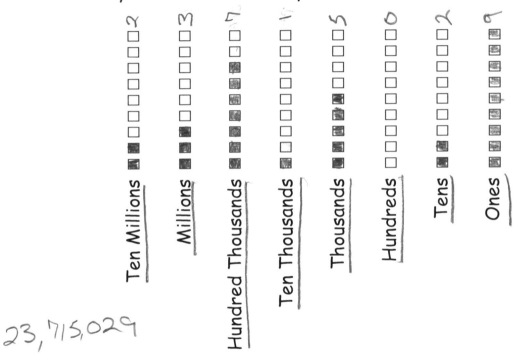

23,715,029

Each place value is 10 times greater than the place value on its right. See that ten 1s make 10, ten 10s make 100, and ten 100s make 1,000!

Each place holder (box) can hold digits 1 through 9, or the column can be empty or zero.

There are only 9 boxes above each place because when 10 is reached, a 1 is added to the next (10 times greater) column to the left, leaving a zero in the current column.

See how the number 1,398,036 can be represented:

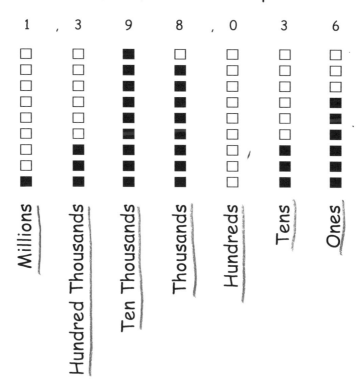

This is the same as writing 1,398,036 in separate pieces:

1,000,000 + 300,000 + 90,000 + 8,000 + 0 + 30 + 6
↑ 1 million   ↑ 3 hundred   ↑ 9 ten   ↑ 8 thousands   ↑ 0 hundreds ↑ 3 tens ↑ 6 ones
                thousands   thousands

Look at another number, 568:

568 = 500 + 60 + 8
        ↑   ↑   ↑
     5 hundreds + 6 tens + 8 ones
     500 ones   + 60 ones + 8 ones

Extra Info: You can think of 568 as
5 hundreds, 6 tens, and 8 ones or just as 568 ones.

In the base ten system:

Ones digit shows how many ones, (1s).

Tens digit shows how many tens, (10s).

Hundreds digit shows how many hundreds, (100s).

Thousands digit shows how many thousands, (1,000s).

Ten Thousands digit shows how many ten thousands, (10,000s).

Hundred Thousands digit shows how many hundred thousands, (100,000s).

Millions digit shows how many millions, (1,000,000s).

And so on...

When you see a large number, remember it can be separated into 1s, 10s, 100s, 1,000s, and so on up as far as you need to go.

Note that **commas** are used in numbers larger than hundreds to separate hundreds and thousands places. Commas also separate hundred-thousands from millions places and every third digit to the left for larger numbers. For example, note the placement of commas in this number with 27 zeros:

1,000,000,000,000,000,000,000,000,000

Note that some countries use a comma in place of the decimal point and use periods, small spaces, or apostrophes in place of commas to segment the digits.

1.3. Even and Odd Integers

Even Integers

   ... -12, -10, -8, -6, -4, -2, 0, 2, 4, 6, 8, 10, 12 ...

Even integers can be divided evenly by 2. So if you have an even number of objects, they can be grouped exactly into pairs.

Let's look! Imagine a group of swimmers. If there is an even number, each swimmer can have one "Buddy" with no swimmers left over.

8 swimmers
4 pairs of swimmers  All swimmers
have Buddies.

As we see with the swimmers, **even integers** can be perfectly divided into pairs, or groups of 2. We can also divide an even number exactly into 2 groups. Let's look again at our 8 swimmers:

8 swimmers
2 groups of swimmers  Swimmers divided
exactly into 2 groups.

Let's do some examples:

Example: How many 2s in 2 faces?

We have 2 faces

2 goes in ⟶   ⟶ 1 time

We see one 2 in our 2 faces. So 2 goes into 2 faces 1 time.

Example: How many 2s in 4 faces?

We have 4 faces

2 goes in ⟶   ⟶ 2 times

We see two 2s in our 4 faces. So 2 goes into 4 faces 2 times.

Example: How many 2s in 6 faces?

We have 6 faces

2 goes in ⟶   ⟶ 3 times

We see three 2s in our 6 faces. So 2 goes into 6 faces 3 times.

Zero is defined as an even integer. Why?

Even integers can be divided evenly, or exactly, by 2.

Zero divides evenly by 2.

Zero divided by 2 gives exactly 0, which is an integer.

Example: How many 2s in 0 faces?

We have 0 faces

2 goes in ⟶ [   ] ⟶ 0 times

So 2 goes into 0 faces 0 times.

Odd Integers

... -9, -7, -5, -3, -1, 1, 3, 5, 7, 9 ...

Odd integers can NOT be divided evenly by 2.
This mean 2 does not divide evenly into odd integers.
If you group an odd number of objects into pairs,
you will have one object left over.

Let's think about the swimmers again. If you have an odd number
of swimmers, then one will be left without a Buddy.

9 swimmers
4 pairs with
1 swimmer left over

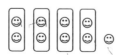

1 swimmer
has no Buddy.

> **Trick**: You can quickly see if a large number is **even** or **odd** by looking to see if the digit in the **ones place** is even or odd.
>
> If **2 will go into the ones digit**, the entire number is even and can be divided evenly, or exactly, by 2.

Consecutive Even Integers

$$\{-12, -10, -8, -6, -4, 2, 0, 2, 4, 6, 8, 10, 12\}$$

Consecutive even integers are even integers arranged in increasing size without any even integers missing in between.

It's like counting by two's!

Consecutive Odd Integers

$$\{-13, -11, -9, -7, -5, -3, -1, 1, 3, 5, 7, 9, 11, 13\}$$

Consecutive odd integers are odd integers arranged in increasing size without any odd integers missing in between.

> Extra info: Fractions are not even or odd.
> The idea of even and odd applies only to integers.

1.4. Practice Problems

1.1

(a) Draw a number line showing only even numbers starting at 22 and ending at 46.

(b) If a number line represents years, where would you find July 4, 1776?

1750 — 1800   1776 — 1777   2015

(c) Why do we put arrows on the ends of number lines?

1.2

(a) Darken the squares in the drawing at the beginning of Section 1.2 to show the number 23,715,029.

(b) If you had thirty $10 dollar bills, how many $100 dollar bills could you get at the bank in exchange?

(c) If you had thirty $10 dollar bills, how many $1 dollar bills could you get?

(d) When you write three zeros to the right of an integer, how much larger is the resulting number?

1.3

(a) If it is important for everyone at your summer camp to have a swimming buddy, how would combining two classes, each with an odd numbers of swimmers, solve the problem?

(b) What if three or five classes each with odd numbers of swimmers were combined?

(c) How could you split a class of 10 swimmers into two groups, each with an even number? How could you split the class of 10 swimmers into two groups, each with an odd number of swimmers?

Answers to Chapter 1 Practice Problems

1.1

(a)

(b) Fractions of years would appear between the year hash marks. Since July 4 is just past half of a year (the 185th day out of 365), it would be located just past half way between the points representing 1776 and 1777.

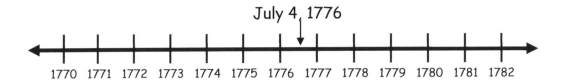

(c) The arrows show that there is an endless, or infinite, number of numbers in both the positive and negative directions.

1.2

(a) The number 23,715,029:

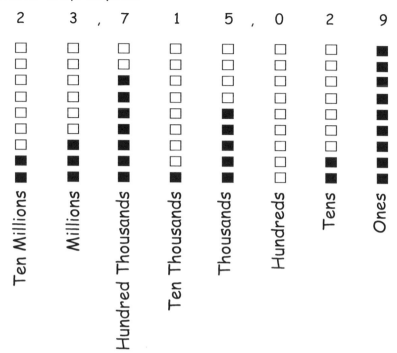

(b) You would receive one $100 bill for every ten $10 dollar bills. Since you had 30 $10 dollar bills, you would receive 3 $100 bills.

$100	**$100**	**$100**
⇕	⇕	⇕
$10 $10 $10 $10 $10 $10 $10 $10 $10 $10	$10 $10 $10 $10 $10 $10 $10 $10 $10 $10	$10 $10 $10 $10 $10 $10 $10 $10 $10 $10

(c) You would receive 10 $1 bills for each $10 dollar bill. If you had 30 $10 dollar bills, you would get 300 $1 dollar bills.

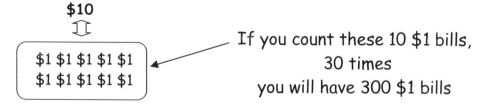

If you count these 10 $1 bills,
30 times
you will have 300 $1 bills

(d) One zero would make the number ten times larger. Two zeros would make the number 10 x 10 or 100 times larger. Three zeros would make the number 10 x 10 x 10 or 1,000 times larger.

1.3

(a) By combining the classes, the extra or "odd" swimmer in one class could become the buddy of the extra swimmer in the other class. This shows how combining or adding two odd numbers results in an even number.

(b) By combining an odd number of classes, each with an odd number of swimmers, you could pair up the extra swimmers in all but one of the classes. This means that the total number of swimmers would be odd.

(c) The group of 10 could be split into groups of 2 and 8 or groups of 4 and 6, which have even numbers. To make groups with odd numbers, the 10 could be divided into two groups of 1 and 9, 3 and 7, or 5 and 5.

Chapter 2
Addition and Subtraction

*But seek ye first the kingdom of God, and his righteousness; and all these things shall be **add**ed unto you. Matthew 6:33*

Simple addition is just like counting.

Kawsay is age 3. After 4 more birthdays, how old will she be?

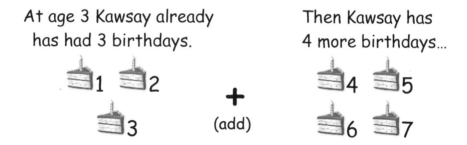

At age 3 Kawsay already has had 3 birthdays.

Then Kawsay has 4 more birthdays...

+
(add)

After 4 more birthdays Kawsay will be age 7.

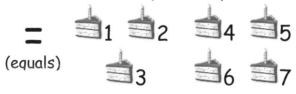

=
(equals)

You can count the seven cakes or else remember that 3 + 4 = 7.

2.1. Add Positive Numbers

Addition means you **add** or combine numbers or amounts.
The **answer**, or total, is called the "**sum**".

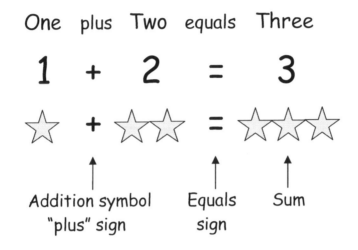

One plus Two equals Three

$$1 \ + \ 2 \ = \ 3$$

Addition symbol   Equals   Sum
"plus" sign   sign

Seeing Addition on the Number Line

You can see addition on the **number line**.
To add a positive number, begin at first number and move right
by the amount of the second number.

Example Add: 1 + 2 = 3

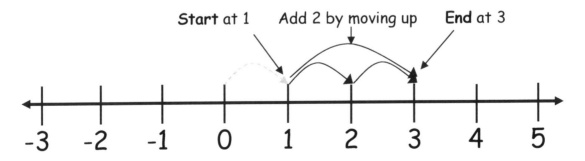

Start at 1   Add 2 by moving up   End at 3

Example: Add 3 plus 5.

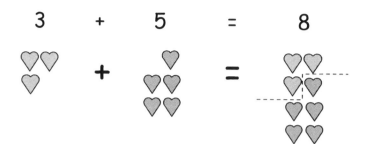

You can see **3 + 5 = 8** on a number line:

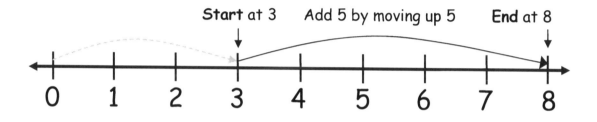

You Can Add 3 or 4 or More Numbers Together

Example: You have 8 marbles. One friend has 6 marbles, and another friend has 2 marbles. How many do you have together?

Let's add the marbles and show addition on the number line:

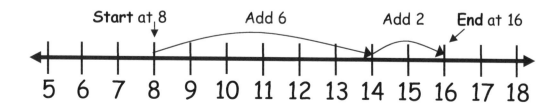

Addition is Commutative and Associative

"Commutative" means the order you add doesn't matter.

For example, we can write:

2 + 3 = 5   and   3 + 2 = 5   They both equal 5.

Example: Show on a number line that 5 + 3 = 3 + 5 = 8.

We first see **5 + 3 = 8** then we see **3 + 5 = 8**

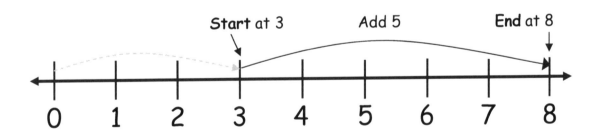

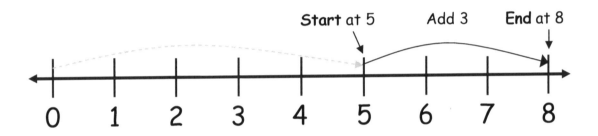

"Associative" means when you add numbers,
it doesn't matter how the numbers are grouped.

For example, 10 marbles can be divided into 3 smaller groups, but
will still add up to 10 marbles.

2 + 3 + 5 = 10   and   2 + 3 + 5 = 10.   They both equal 10.

It doesn't matter if you **first add 2 + 3** or **first add 3 + 5**.

Example: Show that (1 + 2) + 3 = 1 + (2 + 3) = 6.

1 + 2 + 3 = 6   1 + 2 + 3 = 6

   3 + 3 = 6   1 + 5   = 6

Addition Using Columns

Numbers are often added by putting them into columns with ones over ones, tens over tens, hundreds over hundreds, etc. Let's look at doing addition using the **column form**:

Add 5 + 3:

    5 ones
  + 3 ones   or
    8 ones

5 ones + 3 ones = 8 ones

The sum is written below the line.

Add 8 + 7:

    8 ones
  + 7 ones   or
  15 ones

8 ones + 7 ones = 15 ones

15 ones = 1 ten and 5 ones

Add 10 + 20:

   10
 + 20
  30

This can also be written

  1 ten
+ 2 tens
  3 tens
(which is 30 ones)

Add 200 + 100:

  200
 + 100
  300

This can also be written

  2 hundreds
+ 1 hundred
  3 hundreds
(which is 300 ones)

Numbers larger than 10 are often added in columns.

We begin with the ones column on the right. Let's see:

Example: Add 63 + 52 = ?

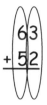

First add ones column.
Then add tens column.

First add **ones column**:  3 ones + 2 ones = 5 ones

♥♥♥   ♥♥   ♥♥♥♥♥

63
+ 52
   5

Then add **tens column**: 6 tens + 5 tens = 11 tens

10 tens + 1 ten

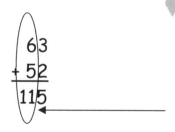

63
+ 52
115

In the answer 115, the left-hand 1 was "carried" into the hundreds column because 10 tens equals 1 hundred.

The answer is: 63 + 52 = 115.

Note: 115 = 1 hundred + 1 ten + 5 ones

Example: Add 89 + 78 = ?

Add in columns:

Add ones column, then add tens column.
Begin adding in the right-hand column and
then move to next column to left.
If the sum in a column is 10 or more, you must
"carry" that left-hand digit to the next column.

First add **ones column**: 9 ones   +   8 ones   =   17 ones

10 ones   7 ones

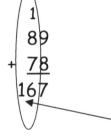

Adding the ones column gives 17 ones.
17 ones = 1 ten + 7 ones. Write the 7 below and
"carry" the 1 ten to the top of the tens column
before adding the tens column.

Add **tens column**:   1 ten + 8 tens +   7 tens =   16 tens

10 tens   6 tens

Adding the tens column gives 16 tens.
16 tens = 1 hundred + 6 tens
The 1 hundred is placed in the hundreds column.

The answer is: 89 + 78 = 167.

Note: 167 = 1 hundred + 6 tens + 7 ones

Example: Add 25 + 36 = ?

```
    1
   25
 + 36
   61
```

We can also write this problem as:

```
   2 tens + 5 ones
 + 3 tens + 6 ones
   5 tens + 11 ones
```

= 5 tens + 1 ten + 1 one

=   6 tens   + 1 one = 61

Therefore, 25 + 36 = 61

Example: Add 59 + 41 = ?

```
   1
   59
 + 41
    0
```

First add ones column: 9 + 1 = 10
Carry the 1.

```
   1
   59
 + 41
  100
```

Next add tens column: 1 + 5 + 4 = 10
Write the 1 in the hundreds column.

Therefore, 59 + 41 = 100

Now let's add three numbers using columns.

Example: Add 68 + 95 + 89 = ?

$$6 \text{ tens} + 8 \text{ ones}$$
$$+9 \text{ tens} + 5 \text{ ones}$$
$$\underline{+8 \text{ tens} + 9 \text{ ones}}$$
$$23 \text{ tens} + 22 \text{ ones}$$

= 2 hundreds + 3 tens + 2 tens + 2 ones

=   200   + 30 + 20 +   2

=   200   +   50   +   2 = 252

Therefore, 68 + 95 + 89 = 252

It is quicker to put in columns and use "carrying". Let's see:

Example: Add 27 + 38 + 49 = ?

(Note the 2 tens that are "carried" to the tens column)

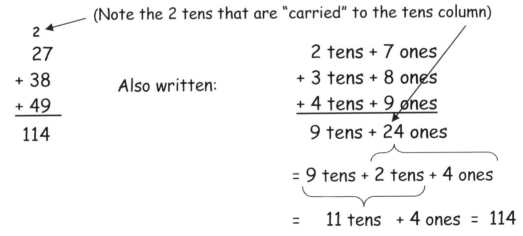

```
  2
  27
+ 38
+ 49
─────
 114
```

Also written:

2 tens + 7 ones
+ 3 tens + 8 ones
+ 4 tens + 9 ones
9 tens + 24 ones

= 9 tens + 2 tens + 4 ones

=   11 tens + 4 ones = 114

Therefore, 27 + 38 + 49 = 114

To add larger numbers, use the same column process!

Example: Add 895 + 78 = ?

(Carry a 10 from ones column and a 100 from tens column)

```
  1 1
  895
+  78
  973
```

We can also write this out as:

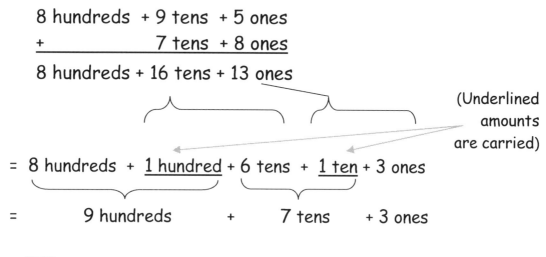

```
  8 hundreds  + 9 tens  + 5 ones
+               7 tens  + 8 ones
  8 hundreds + 16 tens + 13 ones
```

(Underlined amounts are carried)

= 8 hundreds + 1 hundred + 6 tens + 1 ten + 3 ones

=   9 hundreds   +   7 tens   + 3 ones

= 973

Remember: Each column can only hold up to the value 9.
If there are 10 or more, the left-hand number must be
carried to the next column to the left.

2.2. Subtract Positive Numbers and Add Negative Numbers

What is Subtraction?

Subtraction is the opposite or reverse of addition.

Subtract means to take away, deduct, or reduce.

Add means increase, combine, or enlarge.

Subtraction finds the difference between two numbers.

The **symbol** for subtraction is –

Subtracting a positive number is the same as adding a negative number:

Subtract a positive number	is the same as	Add a negative number
5 - 3 = 2	is the same as	5 + (-3) = 2
22 - 10 = 12	is the same as	22 + (-10) = 12
2 - 10 = -8	is the same as	2 + (-10) = -8

"Adding a negative" is just like "subtracting"

Let's look at a number line:

To add a positive number, move right ⟶

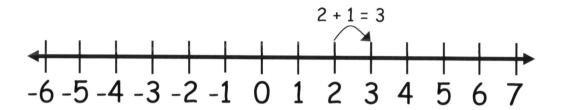

$$2 + 1 = 3$$

-6 -5 -4 -3 -2 -1 0 1 2 3 4 5 6 7

To subtract or add a negative number, move left ⟵

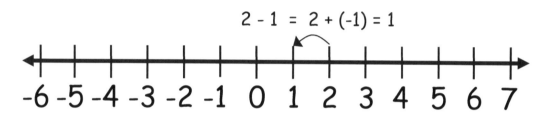

$$2 - 1 = 2 + (-1) = 1$$

-6 -5 -4 -3 -2 -1 0 1 2 3 4 5 6 7

The first negative "-" sign you see means go left

Now let's do an example. Remember, when we add a positive number on the number line, we begin at the first number and move **right**. We again see that as we **subtract a positive number** (which is the same as adding a negative number) we move to the **left**.

Example: Show 5 − 3 = 2

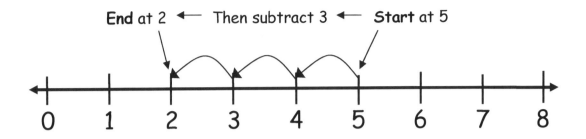

Remember, **5 − 3 = 2** is the same as **5 + (−3) = 2**

Subtraction finds the difference between two numbers.
The answer to a subtraction problem is called the **"difference"**.

Example: Show that the difference between 9 and 7 is 2.

Show the difference **9 − 7 = 2** on the number line:

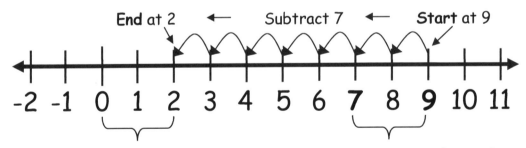

See that the **"difference"** or distance between 9 and 7 is 2.

Next we'll show that subtracting a positive number is the same as adding a negative number using a piggy bank...

Suppose **you have $100 cash in your pig.**
If you remove $10, how much money is left in the pig?

You subtract a positive $10 dollars from $100. You write:

$100   remove $10   → $90

or

$$\$100 - \$10 = \$90$$

Let's start again: Suppose **you have $100 cash in your pig.**
You get an invoice (a debit) in the mail for $10. The invoice
represents money you owe. It is "negative money". If you add
the $10 invoice to the pig, what is the value of money in the pig?

You add a debit of $10 dollars to $100. You write:

$100   add $10 debit   → $90

or

$$\$100 + (-\$10) = \$90$$

Adding an invoice for $10 dollars represents adding a piece of
paper worth negative $10. You "added" something with **negative**
value. The result is the same as removing or subtracting $10.

Here's one more way to think about adding a negative number.

Suppose you are on a highway and the milepost sign says 77 miles. If you turn your car around and drive in the **opposite (negative) direction** for 3 miles, where will you be?

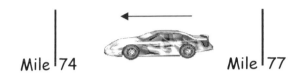

Mile 74   Mile 77

milepost 77 – 3 miles = milepost 74

You have added +3 miles to your car, but your milepost sign change is from 77 back to 74, which is negative 3 miles:

$$77 - 3 = 74$$

You subtract the positive 3 miles that you drove.

Instead of turning around, suppose you keep your car pointing in the positive direction but **drive it in reverse** for 3 miles! So you drive negative 3 miles.

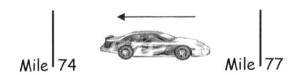

Mile 74   Mile 77

milepost 77 + (–3 miles) = milepost 74

Your car drove backward (in reverse) from Mile 77 to Mile 74. Your milepost location decrease by 3 miles:

$$77 + (-3) = 74$$

You added the negative 3 miles.

Can you show how to add a negative number?

We can show this on the number line.

Example: Show –3 + 5 = ?

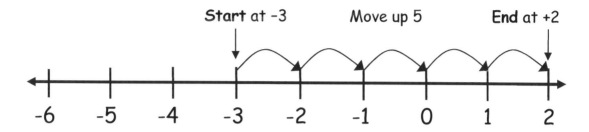

We see that: **–3 + 5 = 2.**

Since the order we do addition doesn't matter:

(–3) + 5 = ? can be written **5 + (–3) = ?** or **5 – 3 = ?**

On the number line, **5 – 3 = ?** looks like:

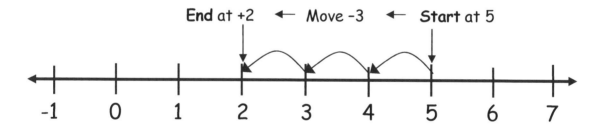

We see that **5 – 3 = 2.**

So, 5 – 3 = 2 and –3 + 5 = 2 and 5 + (–3) = 2.

Adding a negative number is the same as subtracting a positive number for larger numbers too!

Example: −58 + 68 = ?

We see that: 68 + (−58) = 68 − 58 = 10

Subtracting More Than Two Numbers

To **subtract more than two numbers**, you can subtract the first two numbers, then subtract the third number from the difference of the first two, and so on. What is important is to make sure the numbers **preceded by** a "+" sign are added and the numbers **preceded by** a "-" sign are subtracted.

Example: Subtract 15 − 4 − 7 = ?

In this example, 15 has no sign shown, which implies it is +15.

First perform: 15 − 4 = 11

Then perform: 11 − 7 = 4

Therefore, **15 − 4 − 7 = 4**

We can see this on the number line:

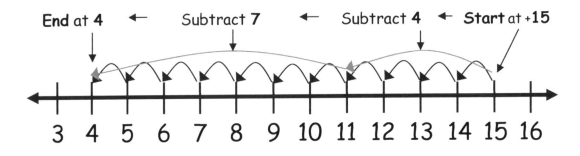

For a **string of additions and subtractions**, you may perform each operation based on the + or - sign preceding each number.

Example: Find 14 – 3 + 7 – 4 - 3 = ?

First perform: 14 – 3 = 11

Next: 11 + 7 = 18

Next: 18 – 4 = 14

Then: 14 – 3 = 11

We see that 14 – 3 + 7 – 4 - 3 = 11.

Let's see this on the number line:

Start at 14 and Subtract 3
Add 7
Subtract 4
Subtract 3 and end at 11

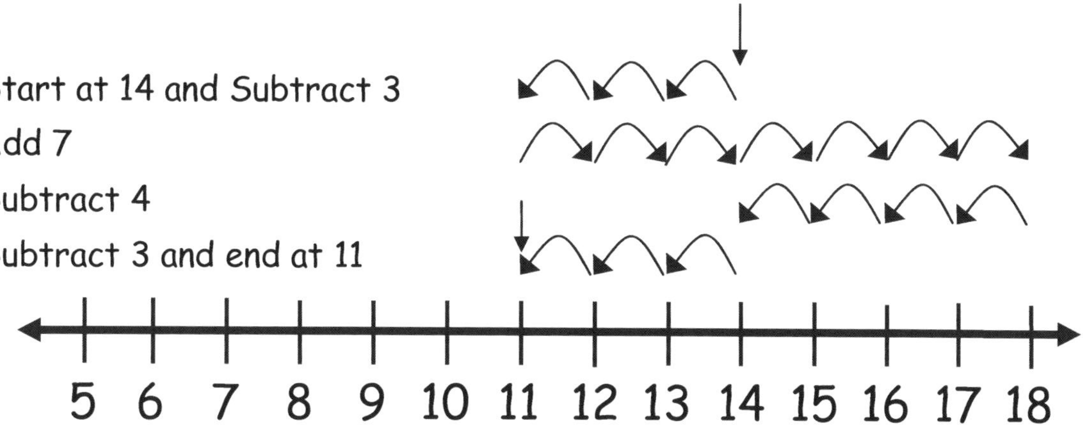

Again, we see that 14 – 3 + 7 – 4 - 3 = 11.

What If There Are Parentheses Present?

If you see additions and subtractions with parentheses, you can first combine the numbers inside the parentheses.
Then add or subtract the numbers outside the parentheses with the combined numbers from inside the parentheses.
Make sure the numbers preceded by a "+" sign are added and the numbers preceded by a "-" sign are subtracted.

Example: Find 14 – (3 + 7) – (4 – 3) = ?

First perform the operations inside the parentheses:

$$(3 + 7) = 10 \text{ and } (4 – 3) = 1$$

 Next put these back into the equation:

$$14 – (10) – (1) = ? \text{ or just } 14 - 10 - 1 = ?$$

You can now add and subtract in order:

First: 14 – 10 = 4

Then: 4 – 1 = 3

We see that 14 – (3 + 7) – (4 – 3) = 3.

Subtracting Larger Numbers Using Columns

To subtract numbers with two or more digits, it is easier to write the numbers in a column format with ones-over-ones, tens-over-tens, hundreds-over-hundreds, etc.
Then you **subtract each column beginning with the ones.**

Just as with addition, you **begin with the right-hand column,** and then proceed to the left.

Example: 65 – 44 = ?

First subtract the ones:
  5 – 4 = 1
Next subtract the tens:
  6 – 4 = 2
So, 65 – 44 = 21

Example: 4,687 – 3,283 = ?

First subtract the ones: 7 – 3 = 4
Next subtract the tens: 8 – 8 = 0
Subtract the hundreds: 6 – 2 = 4
Subtract the thousands: 4 – 3 = 1
So, 4,687 – 3,283 = 1,404

Note: If you subtract a number from itself, the answer is zero

  8 – 8 = 0   The difference between 8 and 8 is zero!

Borrowing

When you are subtracting, "borrowing" is what you do if the bottom number in a column is larger than the top number.

To subtract a larger number (9) from a smaller number above it (5) in the same column, you **"borrow"** one **10** from the number to the left of the 5. So we borrow 1 ten from the 4 tens, leaving 3 tens. (This is the opposite of "carrying" used in addition.)

When you **borrow**, you take one unit from the number to the left (which is 10 times greater) and add that unit to the number that is too small.

$$3\{15\}$$
$$\underline{-\ \ \ 9}$$

When you move 1 ten from the tens column to the ones column, it becomes a 10 because: **1 ten = 10 ones.**
Similarly, 1 hundred becomes 10 tens, and 1 thousand becomes 10 hundreds, etc. If several columns have larger numbers on the bottom, you work from right to left, borrowing from the leftward column if necessary.

Example: 25 - 9 = ?

2|5
-|9

First subtract the **ones column**.
But the 9 ones is larger than the 5 ones.
We must **borrow** from the 2 tens.

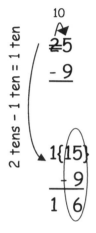

2 tens – 1 ten = 1 ten

10
2̶5
- 9

1 ten is "borrowed" from the 2 tens and
added to the 5 ones: 5 + 10 = 15 ones.
This leaves 2 - 1 = 1 in the tens column.

1{15}
+ 9
1 6

Now we subtract the ones: 15 - 9 = 6
And then subtract the tens: 1 - 0 = 1
Therefore, 25 - 9 = 16

To check subtraction, reverse it by adding the answer, or
difference, to the number that was subtracted: 16 + 9 = 25
Note that carrying the 1 reverses the borrowing.

Note that 25 - 9 = ? can also be written:

	which is	
2 tens + 5 ones		25
– 0 tens + 9 ones	the same as	– 9

Let's solve it again using a different format:

First, we **borrow** 1 ten from the tens column:

1 ten + 15 ones   which is   1{15}
- 0 tens + 9 ones   the same as   - 9

Second, we subtract the ones: 15 ones – 9 ones = 6 ones

 1 tens + 15 ones   which is   1{15}
- 0 tens + 9 ones   the same as   - 9
          6 ones   6

Third, we subtract the tens: 1 ten – 0 tens = 1 ten

 1 tens + 15 ones   which is   1{15}
- 0 tens + 9 ones   the same as   - 9
 1 ten + 6 ones   1 6

Again we see, 25 - 9 = 16, or 1 ten and 6 ones, which is 16.

We can also see **25 - 9 = 16** on the number line:

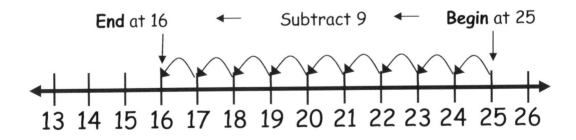

Example: 3,352 - 1,456 = ?

$\begin{array}{r} 3,35\cancel{2} \\ -\ 1,45\cancel{6} \\ \hline \end{array}$ In the ones column, 6 is greater than 2.
You need to borrow from the 5 tens.

$\overset{10}{\curvearrowright}$
$\begin{array}{r} 3,34\{12\} \\ -1,45\ \ 6 \\ \hline \end{array}$ Borrow 1 ten from the 5 tens leaving 4 tens.
This increases the 2 ones to 12 ones.

Oops! Now, in the tens place, the 4 is less than the 5 below it. So you need to borrow from the 3 in the hundreds place.

$\overset{100}{\curvearrowright}$
$\begin{array}{r} 3,2\{14\}\{12\} \\ -1,4\ \ 5\ \ 6 \\ \hline \end{array}$ Borrow 1 hundred from 3 hundreds leaving 2
hundreds. This increases 4 tens to 14 tens.

In the hundreds place, the 2 is less than the 4 below it.

$\overset{1000}{\curvearrowright}$
$\begin{array}{r} 2,\{12\}\{14\}\{12\} \\ -1,\ \ 4\ \ 5\ \ 6 \\ \hline \end{array}$ Borrow 1 thousand from 3 thousands leaving 2
thousands to raise 2 hundreds to 12 hundreds.

Finally, we have set this up so we can do 4 simple subtractions:

$\begin{array}{r} 2,\{12\}\{14\}\{12\} \\ -1,\ 4\ \ \ 5\ \ \ 6 \\ \hline 1,\ 8\ \ \ 9\ \ \ 6 \end{array}$

Ones: 12 - 6 = 6
Tens: 14 - 5 = 9
Hundreds: 12 - 4 = 8
Thousands: 2 - 1 = 1

Our answer is: 3,352 - 1,456 = 1,896.
Check answer: 1,896 + 1,456 = 3,352.

What if you go to borrow from the column to the left and find nothing there but a zero? You just go one additional column to the left and start borrowing!

Example: 2,002 - 987 = ?

2,00<s>2</s>
- 98<s>7</s>

In the ones column, 7 is greater than 2. You can't borrow from the 10s or the 100s, so you have to borrow from the 1,000s.

1000
1,{10}02
-   9 87

Borrow 1 from the thousands to make 10 hundreds. (This leaves 1 thousand.)

100
1,{9}{10}2
-   9 8 7

Borrow 1 from the 10 hundreds to make 10 tens. (This leaves 9 hundreds.)

10
1,{9}{9}{12}
-   9 8 7
1, 0 1 5

Borrow 1 from the 10 tens to make 10 more ones, giving 12 ones. (This leaves 9 tens.) Subtract columns: 12-7, 9-8, 9-9, 1-0.

Our answer is: 2,002 - 987 = 1,015.

Check answer: 1,015 + 987 = 2,002.

Subtracting a Larger Number from a Smaller Number

What if you need to subtract a number with a greater value from a number with a smaller value? In other words, what happens when you see:

Small Number - Big Number = ?

Placing the numbers in columns won't help, since the entire number you are subtracting from is smaller (so there is nothing to "borrow" from). If we look at subtracting a larger number from a smaller number, you will see that you get a negative number. In other words:

Small Number - Big Number = Negative Number

Let's see: Imagine you're in an elevator on the 3rd floor and you go down 4 floors. Where are you? In the basement at -1 floor!

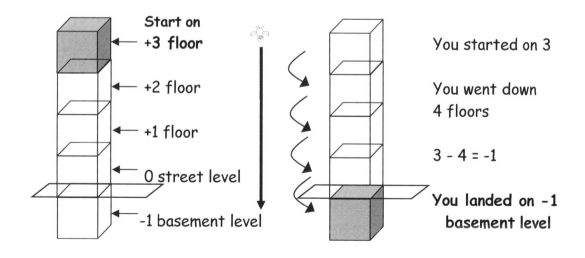

Using an elevator going down 4 floors we see that: 3 - 4 = -1

**Here's another way to look at subtracting
a larger number from a smaller number.**

Suppose you are at the store. You have $10 in your pocket.

But you want to buy a book that costs $15.

Suppose the store owner let's you take home the book today if

you will pay him the rest tomorrow.

How much do you owe the store?

You owe $5, which means you have a $5 debit.

You now have -$5 in your pocket!

Let's look:

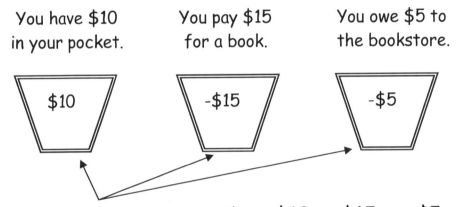

Using your **pocket** we can see that: **$10 - $15 = -$5**

Let's look at $10 - $15 = -$5 on a number line:

10 – 15 = -5

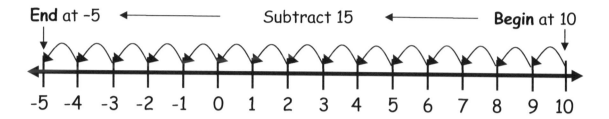

See that we moved through zero on the number line into negative territory.

Example: Subtract larger number 7 from smaller number 3 using a number line.

3 – 7 = ?

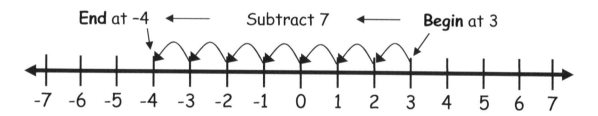

Again, we keep moving to the left through zero on the number line into **negative territory!** We see on the number line that:

3 - 7 = -4   which is a negative number.

There is a Trick
to Subtracting a Larger Number From a Smaller Number

To subtract a large number from a small number:
Reverse! Subtract the small number from the large number.
Then take the opposite or negative.

Let's see:

To calculate:

Small Number - Large Number = ?

Reverse. Calculate. Take the opposite:

Large Number - Small Number = Answer

Take the opposite by writing a negative sign:

– Answer

Why does this trick work? Because subtraction measures the
"**absolute value**" of the difference between two numbers.
The **absolute value is the distance from zero.** This means the:
 absolute value of -7 is 7 and the absolute value of +7 is 7

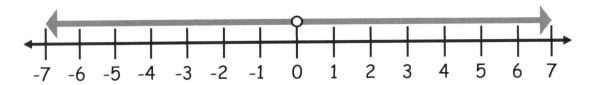

For two numbers, whether you subtract the larger number from the smaller number or the smaller number from the larger number, the absolute value of the answer is the same.

Therefore, 3 - 7 and 7 - 3 have the same absolute value of 4.
So:
Subtracting smaller number 3 from larger number 7 is: 7 - 3 = **4**
Subtracting larger number 7 from smaller number 3 is: 3 - 7 = **-4**

Example: Show 7 - 3 and 3 - 7 on a number line.

7 - 3 = 4 on a number line is **+4** to the right of 0:

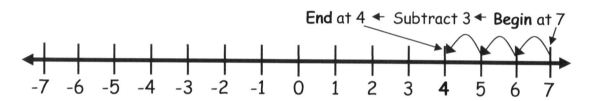

3 - 7 = -4 on a number line is **-4** to the left of 0:

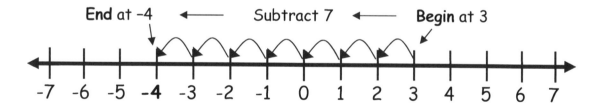

The answer to both 3 - 7 = -4 and 7 - 3 = 4 is absolute value 4. Both are a distance 4 from 0 on the number line, but in opposite directions.

Example: What is 3 – 9 = ?

To subtract a larger number from a smaller number, you can reverse and subtract the small number from the large number. Then take the negative.

So you first find the answer to 9 – 3 = ?, which is 9 - 3 = 6

You can show **9 - 3 = 6**

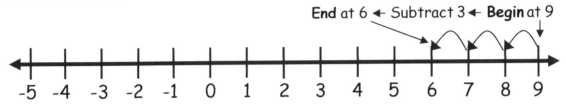

Now you take the negative of answer 6, which is **-6.**

Therefore, since 9 – 3 = 6, then 3 – 9 = -6.

You can also directly show 3 - 9 = -6 on the number line:

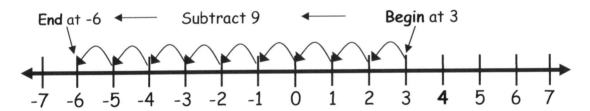

We see that:
Subtracting small number 3 from large number 9 is: 9 - 3 = **6**
Subtracting large number 9 from small number 3 is: 3 - 9 = **-6**

We have again shown that to find the answer to subtracting a large number from a small number, we just subtract the small number from the large number, and then take the negative.

Example: What is 2 – 4 = ?

First subtract 4 - 2 = 2. Then take the negative: -2. Therefore, **2 - 4 = -2.**

Let's check this by looking at an elevator. As we see below: If the elevator moves up from the -2 basement level to the +2 floor, it goes up (positive direction) 4 floors.
But if the elevator moves down from the +2 floor to the -2 basement level, it goes down (negative direction) 4 floors.

We can subtract even when the second number is bigger: 2 - 4 = -2

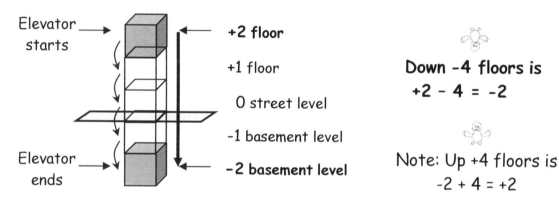

Elevator starts → +2 floor

+1 floor

0 street level

-1 basement level

Elevator ends → -2 basement level

Down -4 floors is +2 – 4 = -2

Note: Up +4 floors is -2 + 4 = +2

What about bigger numbers?

Example: What is 53 – 79 = ?

Again, you reverse, subtract, and take the negative. So subtract small number 53 from large number 79 in a column:

```
    79
   -53
    26
```

Then take the negative: -26. **Therefore, 53 – 79 = -26.**

2.3. Subtract Negative Numbers

What does it mean to subtract a negative number?

Subtracting a negative number is the same as adding a positive number. **Subtracting –7 is the same as adding +7:**

-(-7) = +7. In other words, **minus minus equals plus:**

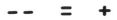

If you have $100 cash and an invoice for a $10 debit in your piggy bank, the total "value" in the pig is $100 + (-$10) = $90.

$100 + (-$10)
= $90

You add a $10 debit to $100:
$100 + (-$10) = $90

Next, what happens to the value of your pig if you remove (subtract) the $10 debit invoice?

$100 + (-$10)
– (-$10) = $100

First you added a $10 debt: **$100 + (-$10) = $90**

Then you subtract the debt: **$90 – (-$10) = $90 + $10 = $100**

When you remove the $10 debit invoice, the pig's value increases by $10. This gets you back to the $100 cash you started with.

Taking away the negative $10 is the same as adding a positive $10! **The negative of a negative is a positive.**

Every time you see a negative or minus sign (-),
think of taking the opposite or negative.

In arithmetic, the negative of a negative is positive.

On the **number line** we saw that both **subtracting a positive** or
adding a negative means moving to the **left**.

Subtracting a 2nd negative means going **opposite of subtracting,**
which means **moving to the right** (just like adding).

Therefore, **when you subtract a negative** using a **number line**:

The first negative sign (-) says to go left.

But the 2nd negative sign (- -) says to go opposite
and switches that direction back to the right.

In other words:

The first negative sign says "go opposite of positive"
(which is negative).

The second negative sign says "go opposite again",
that is, go opposite of negative (which is positive).

Let's look at some examples.

Example: Show on the number line, 3 – (–5) = ?

1st "–" says go negative, the 2nd "–" says go opposite which is positive.

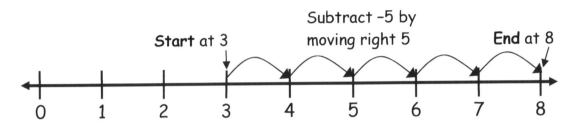

Do you see 3 – (–5) = 8 is just 3 + 5 = 8? It is because -(-5) = +5
This shows that subtracting a negative is the same as adding!
Let's look at a piggy bank:

You begin with **$8 cash plus a $5 "IOU" debit** you owe your dad:

Giving you an initial
value of $3

Dad cancels the $5 IOU debit, so you **subtract** the **-$5 IOU.**

Removing the $5 IOU debit increases your pig's value by $5!
Now your pig is worth $8.

Giving you a final
value of $8

Example: 76 – (–19) = ?

Subtracting a negative is the same as adding!   **– – = +**
You write the double negative as a positive or plus.

$$76 - (-19) = 76 + 19$$

Then add 76 + 19 using columns:

```
  1
 76
+19
 95
```

Therefore, 76 – (–19) = 76 + 19 = 95

Example: 15 – (–22) – 31 + (–2) = ?

Subtracting a negative is the same as adding!   **– – = +**

$$15 - (-22) - 31 + (-2) = 15 + 22 - 31 + (-2) = ?$$

Also, adding a single negative is just subtracting, so rewrite as:

$$15 + 22 - 31 - 2 = ?$$

Finally, add and subtract. (Perform each operation based on the + or - sign preceding each number.)

First: 15 + 22 = 37

Next: 37 – 31 = 6

Next: 6 – 2 = 4

Therefore, 15 – (–22) – 31 + (–2) = 15 + 22 – 31 – 2 = 4

Example: -17 – (-9) = ?

You can write the double negative as a positive or plus:

$$-17 - (-9) = -17 + 9 = ?$$

Let's show -17 + 9 = ? on the number line:

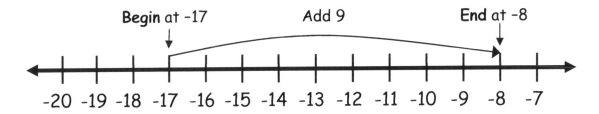

Therefore, we see: -17 – (-9) = -17 + 9 = -8

Here's another way to look at -17 – (-9) = ?

Write the double negative as a positive: -17 – (-9) = -17 + 9

What if you re-order the addition -17 + 9 = ? to:

$$9 + (-17) = ? \quad \text{or just} \quad 9 - 17 = ?$$

Now you can show **9 – 17 = ?** on the number line:

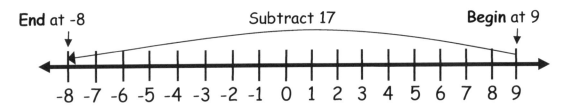

See that: -17 – (-9) = -17 + 9 = 9 – 17 = -8

Remember, we can also find 9 - 17 = ? as:
    the negative of 17 – 9 = 8, which is -8

What If Both Signs Are Negative?

In other words, **what happens when you subtract a negative number from a negative number?**

Example: **−12 − 7 = ?**

We can always look at numbers on the number line. In this case:

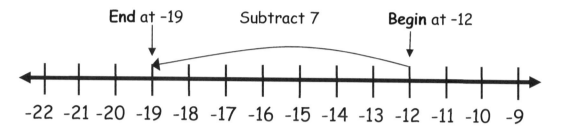

We began at −12, moved 7 in the negative direction, and landed on −19:

 Therefore, −12 − 7 = −19

There is a trick or shortcut to working this type of double negative problem.

We can think of this problem as the **opposite of adding two positives.** That means we can write:

$$-12 - 7 = ? \quad \text{as} \quad 12 + 7 = 19$$

Then take the negative of the answer: −19

Let's apply the trick for subtracting two negative numbers as the opposite of adding two positives to larger numbers.

Example: **–105 – 35 = ?**

Rewrite as the opposite of adding two positives: 105 + 35 = 140
The opposite is -140. **Therefore, –105 – 35 = -140.**

We can see that this works by showing **–105 – 35 = ?** on a number line. Begin at -105 and move in the negative direction (to the left) while you count 35 steps. You land on -140.

–105 – 35 = –140

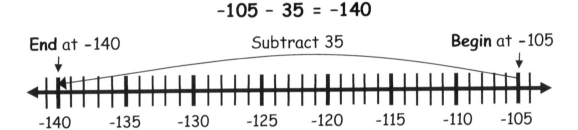

How about more than two negative numbers?

Example: Find -3 - (-4 - (-5 + (-6))) = ?

Start inside the nested
 parentheses and work out!

-3 - (-4 - (-5 + (-6))) =

-3 - (-4 - (-5 - 6)) =

-3 - (-4 - (-11)) =

-3 - (-4 + 11)) =

-3 - (7)) =

-3 - 7 = -10

So, -3 - (-4 - (-5 + (-6))) = -10

> **Extra Credit
> Solution**: You can
> also solve by
> distributing signs into
> the parentheses:
> -3 - (-4 - (-5 + (-6))) =
> -3 - (-4 - (-5 - 6)) =
> -3 - (-4 + 5 + 6) =
> -3 + 4 - 5 - 6 = -10

2.4. Add and Subtract Zero

If zero (nothing) is added or subtracted from a number, the number does not change.

$$Number \; + \; 0 \; = \; Number$$
$$Number \; - \; 0 \; = \; Number$$

If you have 2 bananas and you add or subtract **zero** bananas, you are left with 2 bananas:

If you have an amount of something and you don't add to it or take away from it, you still have the same amount.

Example: $6 - 0 = 6$ and $6 + 0 = 6$

Example: $346 - 0 = 346$ and $346 + 0 = 346$

Example: $x - 0 = x$ and $x + 0 = x$

Example: $0 + 0 = 0$ and $0 - 0 = 0$

2.5. Add and Subtract Even and Odd Integers

In Section 1.3 you learned about even and odd integers and how even integers can be evenly divided by 2.

What if you add or subtract even or odd integers? Will the answer be even or odd?

When you add or subtract, if **both** numbers are even or both numbers are odd, your answer will be an **even** number.

Add or subtract evens gives **even** answer:

6 + 2 = 8   and   6 - 2 = 4

Add or subtract odds gives **even** answer:

7 + 5 = 12   and   7 - 5 = 2

When you add or subtract, if **one** number is even and the other number is odd, your answer will be an **odd** number.

Add or subtract even and odd gives **odd**:

7 + 4 = 11   and   7 - 4 = 3

First see adding and subtracting when the answer is even:

Two even numbers: Remember the swimmers in Section 1.3? Imagine you have two groups. If each group has an even number of swimmers, will everyone within the groups find a buddy? Yes. If you **add** two even groups, will everyone have a buddy? Yes. If you **subtract** two even groups, will the remaining swimmers have buddies? Yes. Let's see:

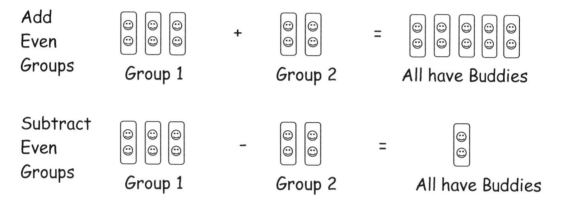

Two odd numbers: If you **add** two odd-numbered groups of swimmers, the extra swimmer from one group can buddy with the extra from the other group, so everyone is paired. This means combining two odd numbers results in a number that is **even**! If you **subtract** two odd-numbered groups, the extra swimmer in the first group is cancelled by the extra swimmer in the second group, leaving no single swimmers, only pairs. Again you have an even number. Let's see:

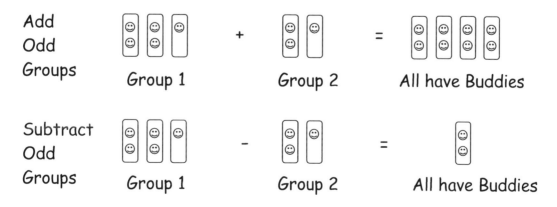

Next see adding and subtracting when the answer is odd:

One even and one odd number: If you combine **one even number** and **one odd number** using addition or subtraction, the result is an **odd number**.

Looking at the swimmers:

If you **add** one **odd group** and one **even group**, there will be one **extra swimmer** without a buddy. You will have an **odd number**. If you **subtract** one **odd group** and one **even group**, there will be one **extra swimmer** left without a buddy. You will have an **odd number**. Let's look:

Add Odd + Even Groups	Group 1	+	Group 2	=	One swimmer has no Buddy. Odd.

Subtract Odd - Even Groups	Group 1	-	Group 2	=	One swimmer has no Buddy. Odd.

Subtract Even - Odd Groups	Group 1	-	Group 2	=	One swimmer has no Buddy. Odd.

In summary, whether you are adding or subtracting, if one number is even and the other is odd, then the answer will always be an **odd** number. Here's a chart combining even and odd. If E means an even number and O means an odd number, then:

$$E + E = E \qquad E - E = E \qquad O + O = E \qquad O - O = E$$

$$E + O = O \qquad E - O = O \qquad O + E = O \qquad O - E = O$$

These results hold whether the answer is positive or negative. Note: if you need to add or subtract a series of numbers, you can do the first two, then that answer with the 3rd, and so on.

Note: In a series of additions or subtractions, the answer will be **odd** if there is an **odd number of odd numbers**! For example:

$3 + 5 + 7 = 15$	Odd number (3) of odd numbers added gives odd answer.
$11 - 5 - 1 = 5$	Odd number (3) of odd numbers subtracted gives odd answer.
$13 + 3 - 1 + 5 - 7 = 13$	Odd number (5) of odd numbers added and subtracted gives odd answer.
$13 + 4 - 2 + 5 - 7 = 13$	Odd number (3) of odd numbers added and subtracted gives odd answer. While there are some even numbers, there is still an odd number of odd numbers (because even numbers don't create unpaired swimmers.)

2.6. Addition and Subtraction Tables

Addition and subtraction tables can help you see and memorize simple additions and subtractions. **To use them, you locate where a number in the left-hand column meets a number in the top row when you move across and down.**

Addition Tables find answers for basic additions, such as:
3 + 2 = ? To use the Table find first-number **3** in the leftmost column. Find second-number **2** in the top horizontal row, and then move down until you are across from first-number **3**. The answer appears where the column and row meet: **3 + 2 = 5** (see dark ovals on Table).

Addition Table

+	0	1	2	3	4	5	6	7	8	9
0	0	1	2	3	4	5	6	7	8	9
1	1	2	3	4	5	6	7	8	9	10
2	2	3	4	5	6	7	8	9	10	11
3	3	4	5	6	7	8	9	10	11	12
4	4	5	6	7	8	9	10	11	12	13
5	5	6	7	8	9	10	11	12	13	14
6	6	7	8	9	10	11	12	13	14	15
7	7	8	9	10	11	12	13	14	15	16
8	8	9	10	11	12	13	14	15	16	17
9	9	10	11	12	13	14	15	16	17	18

Notice in the Addition Table above, if you move diagonally from the lower left to the upper right, the numbers are the same. For example, the 9's are in a dashed rectangle.

Example: Show 4 + 3 in the Table below.

Find 4 in the left column. Then find 3 in the top row and move down until you are across from 4. That box shows a 7.
Therefore, we see: 4 + 3 = 7

Addition Table

+	0	1	2	3	4	5	6	7	8	9
0	0	1	2	3	4	5	6	7	8	9
1	1	2	3	4	5	6	7	8	9	10
2	2	3	4	5	6	7	8	9	10	11
3	3	4	5	6	7	8	9	10	11	12
4	4	5	6	7	8	9	10	11	12	13
5	5	6	7	8	9	10	11	12	13	14
6	6	7	8	9	10	11	12	13	14	15
7	7	8	9	10	11	12	13	14	15	16
8	8	9	10	11	12	13	14	15	16	17
9	9	10	11	12	13	14	15	16	17	18

Note that you can also add 4 + 3, or equivalently 3 + 4, by first finding 3 in the left column, then finding 4 in the top row, and moving down until you are across from 3. That box also shows 7!

Subtraction Tables find answers for basic subtractions. The numbers in the table are the answers you get when you **subtract a top-row number from a left-column number.** For example, to subtract: **7 - 2 = ?** Find the first-number **7** in the leftmost column. Then find second-number **2** from the top row. The answer appears where the column and row meet: **7 - 2 = 5**

Subtraction Table

-	0	1	2	3	4	5	6	7	8	9
0	0	-1	-2	-3	-4	-5	-6	-7	-8	-9
1	1	0	-1	-2	-3	-4	-5	-6	-7	-8
2	2	1	0	-1	-2	-3	-4	-5	-6	-7
3	3	2	1	0	-1	-2	-3	-4	-5	-6
4	4	3	2	1	0	-1	-2	-3	-4	-5
5	5	4	3	2	1	0	-1	-2	-3	-4
6	6	5	4	3	2	1	0	-1	-2	-3
7	7	6	5	4	3	2	1	0	-1	-2
8	8	7	6	5	4	3	2	1	0	-1
9	9	8	7	6	5	4	3	2	1	0

As you move diagonally from the upper left to the lower right the entries stay the same (see 2's in the dashed rectangle).

Example: Subtract 6 - 3 using the Subtraction Table below.

Find **6** in the leftmost column. Then find **3** from the top-row.
The answer to **6 - 3** appears where **column and row meet at 3.**
(**See ovals in Table.**) **Therefore, we see: 6 - 3 = 3.**

Example: Subtract 3 - 5 using the Subtraction Table below.

Find **3** in the leftmost column. Then find **5** from the top-row.
The answer to **3 - 5** appears where **column and row meet at -2.**
(**See rectangles in Table.**) **Therefore, we see: 3 - 5 = -2.**

Subtraction Table

-	0	1	2	3	4	5	6	7	8	9
0	0	-1	-2	-3	-4	-5	-6	-7	-8	-9
1	1	0	-1	-2	-3	-4	-5	-6	-7	-8
2	2	1	0	-1	-2	-3	-4	-5	-6	-7
3	3	2	1	0	-1	-2	-3	-4	-5	-6
4	4	3	2	1	0	-1	-2	-3	-4	-5
5	5	4	3	2	1	0	-1	-2	-3	-4
6	6	5	4	3	2	1	0	-1	-2	-3
7	7	6	5	4	3	2	1	0	-1	-2
8	8	7	6	5	4	3	2	1	0	-1
9	9	8	7	6	5	4	3	2	1	0

In the past students used "**flash cards**" to drill each other on arithmetic. A flash card had the question on one side and the answer on the other. Students memorized simple addition, subtraction, multiplication, and division problems, so they could quickly answer when shown a flash card. The idea was to get to the point where if someone asked you while you slept, "What is 17 minus 8?" you would mumble, "9" without waking up! In general, it is a good idea to memorize the answers to basic addition, subtraction, multiplication, and division calculations.

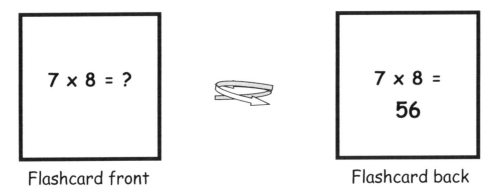

Flashcard front   Flashcard back

Another way you can remember is with **fun rhymes**, such as:

My dog Fido can do tricks!

Seven times eight is **fifty-six**!

Then if you remember a few rhymes you can count up or down. For example, if you remember:

$7 \times 8 = 56$, or eight 7s is 56

Then nine 7s is 7 more than 56, or 63. So $7 \times 9 = 63$

And seven 7s is 7 less than 56, or 49. So $7 \times 7 = 49$

You can also memorize perfect squares such as: 2×2, 3×3, etc.

2.7. Practice Problems

Simple addition is just like counting. Quack!

2.1

(a) You find 4 bags of marbles and dump them into 4 piles:

Bag A: ⬭⬭⬭⬭ Bag B: ⬭⬭⬭ Bag C: ⬭⬭⬭⬭ Bag D: ⬭⬭
     ⬭⬭⬭⬭   ⬭⬭   ⬭⬭⬭   ⬭⬭

Find the total number of marbles by counting them.

(b) In 2.1(a) count the marbles in each pile and write an addition problem which expresses the sum of the 4 numbers.

(c) In 2.1(a) show the marble addition using a number line.

(d) Find the sum of 1,234 and 2,345 using column form.

(e) Find 3,456 + 6,789 using column form.

2.2

(a) You have 26 duck eggs. 8 hatch. How many eggs are left?

Before: ⬭⬭⬭⬭⬭⬭⬭⬭⬭⬭⬭⬭⬭⬭⬭⬭⬭⬭⬭⬭⬭⬭⬭⬭⬭⬭

After: ⬭🐤⬭⬭🐤⬭⬭🐤⬭⬭🐤⬭🐤🐤⬭⬭⬭⬭🐤🐤⬭⬭⬭⬭⬭🐤⬭

(b) Show 27 – 7 – 6 – 9 on a number line.

(c) Show 18 + 7 – 19 + 4 on a number line.

(d) Use column format to solve these 3 problems:
201 – 92 = ?   and   777 – 1,616 = ?   and   –456 + 123 = ?

2.3

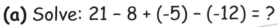

(a) Solve: 21 – 8 + (-5) – (-12) = ?

(b) You have saved $21,015. You spent $3,952 on a vacation. Then your uncle cancel's an IOU you owe him for $5,000. Using subtraction, how much do you have left?

(c) EXTRA CREDIT PROBLEM: Your hot air balloon is tied down and ready to take off. It is pulling up with a force of 200 lbs. You weigh 150 lbs. You step on board and toss one 50 lb weight and one 25 lb weight overboard. Write and solve a subtraction problem to find the new upward force.

2.4

(a) If E means an even number and O means an odd number, is E + O + O going to be even or odd?

(b) If you add three even numbers and six odd numbers together, is the sum even or odd?

(c) Is O – O + E – O – E – O + O even or odd?

2.5

(a) Memorize the addition and subtraction tables until you can recite each row out loud in under 30 seconds. For example:
"4 + 0 = 4. 4 + 1 = 5. 4 + 2 = 6. 4 + 3 = 7. 4 + 4 = 8. 4 + 5 = 9. 4 + 6 = 10. 4 + 7 = 11. 4 + 8 = 12. 4 + 9 = 13."

(b) Construct larger addition and subtraction tables running from 0 to 20 (or more) in each direction and fill in the answers. Look for patterns in the entries.

Answers to Chapter 2 Practice Problems

2.1

(a) You should count 24 marbles.

(b) 8 + 5 + 7 + 4 = ?

(c)

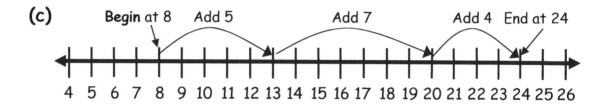

(d)   1,234
           +2,345
           3,579

(e)   1 1 1 1
           3,456 ＼ Carry to the tens, hundreds,
          +6,789   thousands & ten thousands places.
          10,245

2.2

(a) Subtract 26 - 8 = ? using column format:

Start with: 26   Borrow 1 from tens: 1{16}   Subtract columns: 1{16}
            - 8   - 8   - 8
             ?   ?   1 8

Therefore, 18 eggs are left.

(b)

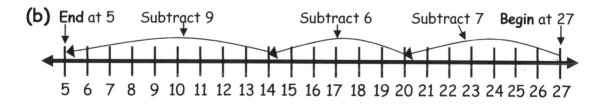

(c)

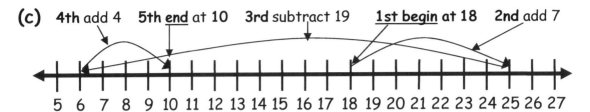

4th add 4   5th end at 10   3rd subtract 19   1st begin at 18   2nd add 7

5 6 7 8 9 10 11 12 13 14 15 16 17 18 19 20 21 22 23 24 25 26 27

(d)   201   First borrow from 100s: 1{10}1   Then borrow from 10s: 1{9}{11}
           -92   - 9 2   - 9 2
            ?   ?   1 0 9

Therefore, 201 – 92 = **109**.

To solve 777 - 1,616 = ? reverse the order, then take the negative:

1,616   First borrow from 10s: 1,60{16} Then borrow from 100s: 1,5{10}{16}
- 777   - 77 7   - 7 7 7
  ?   ?   8 3 9

Taking the negative of the answer gives: **-839**.

Solve -456 + 123 = ? by reversing and using column format:

     123   Reverse the order: 456
    -456   -123
      ?   333

Taking the negative of the answer gives: **-333**.

2.3

(a) Simplify by remembering that adding a negative is like subtracting, and that subtracting a negative is like adding. Then you can solve from left to right, adding when numbers are preceded by + signs and subtracting when numbers are preceded by - signs:

21 – 8 + (-5) – (-12) = ? becomes: 21 – 8 – 5 + 12 = ?

First solve: 21 – 8 = 13, next: 13 – 5 = 8, finally: 8 + 12 = 20

Therefore, 21 – 8 + (-5) – (-12) = **20**.

(b) Spending on your vacation is like subtracting positive dollars. Having your uncle forgive a loan, however, is like subtracting or taking away negative dollars, which is like adding. This results in: $21,015 - $3,952 + $5,000 = ? First subtract using columns:

```
  21,015   Borrow from 100s, 1,000s & 10,000:   1{10},{9}{11}5
 - 3,952                                        -3,  9   5 2
                                                $17, 0  6 3
```

So $21,015 - $3,952 = $17,063.

Next **add** the $5,000 of the forgiven IOU:

$17,063 + $5,000 = **$22,063**

(c) A force upward is in the positive direction. Your weight is a downward or negative force. If you remove weight, you subtract from the downward negative force. The math problem is:

+200 balloon + (-150 your weight) – (-50 toss weight) – (-25 toss weight) = ?

First simplify, remembering that adding a negative is like subtracting, and subtracting a negative is like adding:

200 – 150 + 50 + 25 = ?

Now solve: First 200 – 150 = 50, then
50 + 50 = 100, and finally 100 + 25 = **125 lbs of upward force**.

2.4

(a) E + O + O going left to right: first E + O = O, next O + O = E. The answer is **even**. You can check with simple numbers:
2 + 3 + 3 = 8.

(b) The three even numbers add to another even number. The six odd numbers also add to an even number because there are six "unpaired" ones which can form 3 pairs. Therefore, the sum of the even and odd numbers will be **even**. Note: Unless there is an odd number of odd numbers, the answer will be even.

(c) You can work this left to right a step at a time. To save time, though, since there are an odd number (5) of odd numbers, the answer will be **odd**. To check this substitute simple numbers:
O – O + E – O – E – O + O = 3 – 3 + 2 – 3 – 2 – 3 + 3 = -3.

Chapter 3
Multiplication and Division

*And the Word of God increased; and the number of the disciples **multiplied**...Acts 6:7*

You can SEE multiplying & dividing

Multiplying is like adding the same number over and over.

12 Eggs   multiplied by   3   =   12 + 12 + 12   =   36 Eggs

 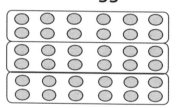

We have 12 eggs 3 times.
Multiplying by 3 combined 3 cartons with 12 eggs each.

Dividing is separating something into a number of smaller parts or groups that are each the same size.

12 Eggs   divided by   3 Eggs   =   4 Groups of 3 Eggs

Dividing by 3, separates the 12 eggs into 4 equal 3-egg groups.

3.1. Multiply

Multiplication is adding the same number over and over again.
It is a shortcut for repeated addition of the same number.

The answer you get from multiplication is called the **product**.

There are many ways to write "multiply", such as: $\times$, * , $\cdot$, ()()

They all mean the same thing. Let's show multiplying the
numbers 5 and 7. We can write 5 times 7 as:

$$5 \times 7 \ = \ 5 * 7 \ = \ 5 \cdot 7 \ = \ (5)(7) \ = \ 35$$

Let's see why we need multiplication!

Imagine you have 7 full cups of sugar and each cup holds 8 oz.
How many total ounces of sugar do you have?

You could add: 8 oz + 8 oz + 8 oz + 8 oz + 8 oz + 8 oz + 8 oz = ?

Let's try adding each of the seven 8 oz together:

8 + 8 = 16

16 + 8 = 24

24 + 8 = 32

32 + 8 = 40

40 + 8 = 48

48 + 8 = 56

The answer is 56, but is there a faster way? Yes: multiplication!

Let's Visualize Multiplication

Counting dots is one way to see multiplication.

Example: How can you see 5 x 4 = ?

5 times 4 is just 5 four times, which is 5 added together 4 times:

$$5 + 5 + 5 + 5 \quad = \quad 20$$

We can count 5 four times to get 20. This shows that adding four 5s is the same as multiplying 5 times 4, or just (5)(4) = 20.

Example: Show 4 x 7 = ?

4 times 7 is just 4 seven times, which is 4 added together 7 times:

$$4 + 4 + 4 + 4 + 4 + 4 + 4 = ?$$

You take the first number and add it as many times as the second number. Let's show it:

1st	2nd	3rd	4th	5th	6th	7th	
4	+ 4	+ 4	+ 4	+ 4	+ 4	+ 4	= 28

The Number Line is Another Way to See Multiplication

In the previous chapter we used number lines to see addition and subtraction. We can also see simple multiplication problems on a number line.

Example: On a number line show 5 x 3, meaning add 5 three times:

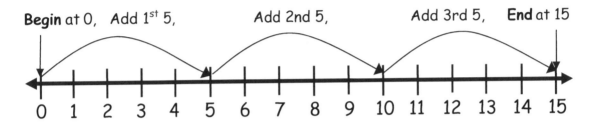

We see that 5 x 3 is just 5 three times!

We also see that: 5 x 3 = 15.

Multiplication Can Be Seen As an Area

In fact, **multiplication of two numbers defines area.**

Example: Show 5 x 5 = ? as an area of dots.

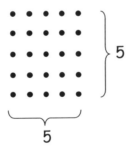

If we count the dots, we will find 25. This shows that 5 x 5 = 25.

We can also visualize the multiplication of two numbers as an area made up of squares. The squares can be square feet, square meters, square miles, or any other length units. For example, a square that is one foot long and one foot wide makes up one **square foot.**

Example: Imagine you have a garden that is 12 feet by 8 feet. How many square feet does it contain?

This is an area question. You can draw out a 12 ft by 8 ft grid of squares. Each square represents 1 square foot. Then you can count the squares. Let's draw the garden and count the squares across:

8 ft

1	2	3	4	5	6	7	8	9	10	11	12
13	14	15	16	17	18	19	20	21	22	23	24
25	26	27	28	29	30	31	32	33	34	35	36
37	38	39	40	41	42	43	44	45	46	47	48
49	50	51	52	53	54	55	56	57	58	59	60
61	62	63	64	65	66	67	68	69	70	71	72
73	74	75	76	77	78	79	80	81	82	83	84
85	86	87	88	89	90	91	92	93	94	95	96

12 ft

By counting each square, you can see that your garden is 96 square feet. In your drawing above, you can also see that your garden is made of 8 horizontal rows. Each row is 1 foot high and 12 feet across. That means each row contains 12 square feet.

Instead of counting each square, another way to find the **area of your garden**, is to add the 12 square feet that make up each row 8 times (since there are 8 rows):

$$12 + 12 + 12 + 12 + 12 + 12 + 12 + 12 = 12 \times 8 = (12)(8) = 96$$

Or, you can find the **garden's area** by counting the 12 columns which are each 8 feet long. You could add 8 twelve times:

$$8 + 8 + 8 + 8 + 8 + 8 + 8 + 8 + 8 + 8 + 8 + 8 = 8 \times 12 = (8)(12) = 96$$

Whether we draw a 12 by 8 grid and count each square or we add eight 12s or we add twelve 8s or we multiply 12 × 8 or we multiply 8 × 12, we find that the garden has 96 square feet!

Multiplication is Commutative and Associative

Commutative means the order we multiply numbers doesn't matter. Remember, the order that we added numbers also did not affect the result. Addition and multiplication are related, since multiplication is adding the same number over and over again. **So the order that numbers are multiplied does not affect the result.** Let's look at an example:

Example: Show 2 x 3 = 3 x 2.

We can show 2 x 3 is the same as 3 x 2 using dots.

```
  3 x 2        equals      2 x 3
 ┌─────────┐              ┌──────┐
 │ • • • │                │ • • │
 └─────────┘              └──────┘
 ┌─────────┐              ┌──────┐
 │ • • • │                │ • • │
 └─────────┘              └──────┘
                          ┌──────┐
                          │ • • │
                          └──────┘
```

It doesn't matter if
we write
2 x 3 or 3 x 2,
we still get 6 dots!

We see 3 x 2 = 2 x 3 = 6 dots.

Let's look at 3 x 2 and 2 x 3 a slightly different way.

3 x 2 is just 3 two times   and   2 x 3 is just 2 three times:

```
  3  +  3   =   6              2  +  2  +  2  =  6
 oo    oo     ooo oo           o     o     o     ooo
  o     o     o o o            o     o     o     ooo
```

We can again see 3 times 2 is the same as 2 times 3:

Whether we write 2 x 3 or 3 x 2, we still get 6 dots or circles!

Example: Show that 2 x 3 = 3 x 2 = 6 on a number line?

We can see the commutative property on a number line:

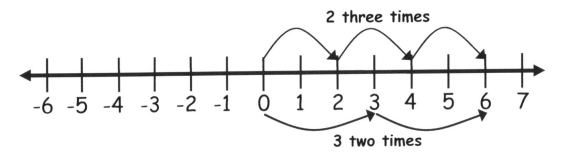

Example: Show 3 x 5 = 5 x 3 using rectangular grid of dots.

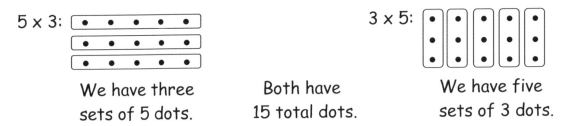

5 x 3:

3 x 5:

We have three
sets of 5 dots.

Both have
15 total dots.

We have five
sets of 3 dots.

Example: Show 4 x 6 = 6 x 4 as grid of dots and circle the sets being added.

4 x 6: 1 2 3 4
        1 2 3 4 5 6

6 x 4: 1 2 3 4 5 6
        1 2 3 4

You can visualize these two groups of dots as the same:
The first one has just been "tipped" over, or rotated, and the
ovals redrawn. In either case, you have 24 dots!

Example: Can a 2 x 4 "tower" fall on its side and become a 4 x 2 slab?

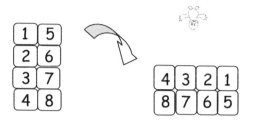

No matter how you look at it, there are exactly 8 blocks in the building!

Example: Is 5 x 7 the same as 7 x 5?

Let's draw it out and see:

Here we see 5 groups of 7 dots. This shows 7 x 5 = 35 dots.

Here is 7 groups of 5 dots. This shows 5 x 7 = 35.

We have seen that the order we multiply doesn't matter!

The associative property of multiplication means it doesn't matter how numbers are grouped when you multiply them:

(2 x 3) x 6 = 36 and 2 x (3 x 6) = 36. They both equal 36.

The **parentheses** mean you multiply the numbers inside first. Then multiply that answer by the number outside. Let's see:

$$(2 \times 3) \times 6 = 36 \quad = \quad 2 \times (3 \times 6) = 36$$

$$6 \times 6 \qquad\qquad 2 \times 18$$

$$36 \qquad = \qquad 36$$

We can draw 2 x (3 x 6) = (2 x 3) x 6 as any of the following:

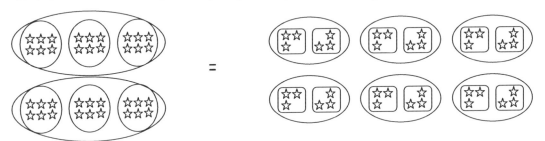

which equals 3 x (2 x 6) or 6 x (3 x 2):

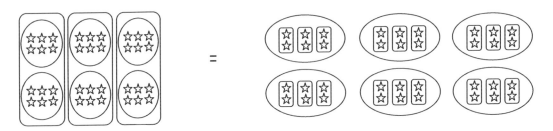

Can you think of other ways to draw these?

Changing the grouping of the factors 2, 3, and 6 does not change the product, or answer, when we multiply.

The **associative property** is consistent with the **commutative property** of multiplication (the order you multiply doesn't matter). The order or grouping of the numbers we multiply doesn't matter.

We can combine associative and commutative and then write:

$$2 \times (3 \times 6) = (2 \times 3) \times 6 = 3 \times (2 \times 6) = (3 \times 2) \times 6$$

$$= 6 \times (3 \times 2) = (6 \times 3) \times 2 = 6 \times (2 \times 3) = (6 \times 2) \times 3$$

We can see that the order or grouping of how we multiply doesn't matter. The answer is still the same.

If you have three numbers you can order them six different ways and group each of them two different ways. You can see there are a total of twelve different ways to multiply them:

$$(3 \times 5) \times 8 = 120$$
$$3 \times (5 \times 8) = 120$$
$$(3 \times 8) \times 5 = 120$$
$$3 \times (8 \times 5) = 120$$
$$(5 \times 3) \times 8 = 120$$
$$5 \times (3 \times 8) = 120$$
$$(5 \times 8) \times 3 = 120$$
$$5 \times (8 \times 3) = 120$$
$$(8 \times 3) \times 5 = 120$$
$$8 \times (3 \times 5) = 120$$
$$(8 \times 5) \times 3 = 120$$
$$8 \times (5 \times 3) = 120$$

What If You Multiply By 1?

Multiplying by 1 just means adding a number one time.

Examples:

$$7 \times 1$$

means adding 7 one time:

$$7 \times 1 \ = \ +7 \ = \ 7$$

$$632,795 \times 1$$

means adding 632,795 one time:

$$632,795 \times 1 \ = \ +632,795 \ = \ 632,795$$

$$0 \times 1$$

means adding 0 one time:

$$0 \times 1 \ = \ +0 \ = \ 0$$

What If You Multiply By 0?

If a number is multiplied by zero, the answer is always **zero**.

If zero is multiplied by any number the result is also **zero**.

number x 0 = 0   or   0 x number = 0

6 x 0 = 0 or 0 x 6 = 0

369 x 0 = 0 or 0 x 369 = 0

9,522 x 0 = 0 or 0 x 9,522 = 0

3,325,000 x 0 = 0 or 0 x 3,325,000 = 0

No matter how many zeros you multiply, you still have only zero!

Remember, multiplication is adding the same number over and over again, so multiplying by 0 means adding a number **zero** times.

7 x 0

means adding 7 zero times:

7 x 0 = 7 zero times = 0

Examples:

9 x 0 means adding 9 zero times.

Zero 9s is just plain zero!

842,569 x 0 means adding 842,569 zero times.

842,569 zero times is just 0.

Let's look at this:

If you have 2 balls times 1, you have 2 x 1 = 2 balls.

This means you have 2 balls **one time**, which is 2 balls.

● ● x 1   =   ● ●

Instead, if you now have 2 balls times 0, you have 2 x 0 = 0 balls.

This means you have 2 balls **zero times**, which is zero balls.

● ● x 0   =   **zero**

When you have something **zero times**, you have it **no times**. You have nothing.

What if We Multiply By Numbers Ending with Zeros?

Remember, multiplying by a certain number such as 7 just means adding the number 7 a certain number of times. For numbers with zeros it still means the same thing.

For example, **7 x 10** means adding 7 ten times:

$$7 \times 10 \; = \; +7+7+7+7+7+7+7+7+7+7 \; = \; 70$$

For **7 x 100** we need to write +7 one hundred times!

$$7 \times 100 \; =$$

+7+7+7+7+7+7+7+7+7+7
+7+7+7+7+7+7+7+7+7+7
+7+7+7+7+7+7+7+7+7+7
+7+7+7+7+7+7+7+7+7+7
+7+7+7+7+7+7+7+7+7+7
+7+7+7+7+7+7+7+7+7+7
+7+7+7+7+7+7+7+7+7+7
+7+7+7+7+7+7+7+7+7+7
+7+7+7+7+7+7+7+7+7+7
+7+7+7+7+7+7+7+7+7+7

$$= \; 700$$

Or for 7 x 1,000 we need to write +7 one thousand times!

The good news is that there is a shortcut to knowing what we get when we multiply by numbers ending with zeros. Let's look…

Let's look at the shortcut to multiplying any number, such as 7, by 10, 100, 1,000, and 10,000:

7 x 10 = 70   Times 10 we put in 1 zero to show 10x.

7 x 100 = 700   Times 100, put in 2 zeros to show 100x.

7 x 1,000 = 7,000   Times 1,000, put in 3 zeros to show 1,000x.

7 x 10,000 = 70,000   Times 10,000, put in 4 zeros to show 10,000x.

Example: Multiply 50 by 10, 100, 1,000, 10,000, and 100,000:

50 x 10 = 500   Times 10, put in 1 zero to show 10x.

50 x 100 = 5,000   Times 100, put in 2 zeros to show 100x.

50 x 1,000 = 50,000   Times 1,000, put in 3 zeros for 1,000x.

50 x 10,000 = 500,000   Times 10,000, put in 4 zeros for 10,000x.

50 x 100,000 = 5,000,000 Times 100,000, put in 5 zeros for 100,000x.

Note: Underlines show the zeros in 10, 100, 1,000, 10,000, 100,000.

Multiplication Table

Multiplication Tables are similar to the Addition and Subtraction Tables in the previous chapter. In a Multiplication Table, **the boldface numbers down the left side are multiplied by the boldface numbers across the top to yield the products contained in the boxes**. For example, we see **3 x 7 = 21**, since row **3** meets column **7** at **21**. (See ovals.)

Multiplication Table

x	0	1	2	3	4	5	6	7	8	9
0	0	0	0	0	0	0	0	0	0	0
1	0	1	2	3	4	5	6	7	8	9
2	0	2	4	6	8	10	12	14	16	18
3	0	3	6	9	12	15	18	21	24	27
4	0	4	8	12	16	20	24	28	32	36
5	0	5	10	15	20	25	30	35	40	45
6	0	6	12	18	24	30	36	42	48	54
7	0	7	14	21	28	35	42	49	56	63
8	0	8	16	24	32	40	48	56	64	72
9	0	9	18	27	36	45	54	63	72	81

Example: Find 8 x 9 = ? in the Multiplication Table.

See dashed ovals above, where the **8**-row meets the **9**-column at **72**. So, **8 x 9 = 72**.

Note that in the Table the **0**-column and the **0**-row show that anything times 0 is 0. Also, the **1**-column and the **1**-row show that 1 times any number is that number. Looking at the **2**-row and the **2**-column, see the numbers going across are the same as the numbers going down. Also notice that the numbers grouped diagonally from upper left to lower right, shown by the diagonal rectangle below, are called "**perfect squares**". These are **numbers times themselves**, such as: **6 x 6 = 36** and **7 x 7 = 49**. Other perfect squares not shown include: **10 x 10 = 100**, **12 x 12 = 144**, and **16 x 16 = 256**.

Multiplication Table

x	0	1	2	3	4	5	6	7	8	9
0	0	0	0	0	0	0	0	0	0	0
1	0	1	2	3	4	5	6	7	8	9
2	0	2	4	6	8	10	12	14	16	18
3	0	3	6	9	12	15	18	21	24	27
4	0	4	8	12	16	20	24	28	32	36
5	0	5	10	15	20	25	30	35	40	45
6	0	6	12	18	24	30	36	42	48	54
7	0	7	14	21	28	35	42	49	56	63
8	0	8	16	24	32	40	48	56	64	72
9	0	9	18	27	36	45	54	63	72	81

It is helpful to memorize the multiplication products shown on the above table! It will help when you multiply larger numbers. Also, try making a 20 by 20 table and filling in the products.

Multiplying Larger Numbers

Large or complex multiplication problems can be broken down into a series of simple multiplications and additions, so all you have to remember is the basic multiplication and addition tables.

Let's first visualize multiplying larger numbers such as 29 x 7? Imagine 7 jars each containing 29 marbles.

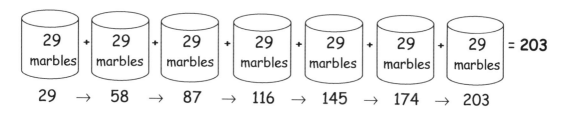

To find out how many marbles we have in all 7 jars combined, we could count them, but that would take a long time! Or we could add 29 seven times. We could also **estimate** the answer by **rounding** the 29 marbles to 30 marbles and then multiplying 7 x 30. Since 7 x 3 = 21, then 7 x 30 = 210 marbles. But our estimate involved an extra 7 marbles, so we can subtract those back off and get 203. While estimating may not quickly lead to the exact answer, it is very useful for checking whether your calculated answer makes sense!

There is, however, an easy way to calculate a multiplication problem using a **column format**. Let's look at that!

Column Format

When you multiply numbers with two or more digits, it is easier to write the numbers in a column format. Then **multiply each digit in the top number (multiplicand), beginning with the ones digit, by each digit in the bottom number (multiplier), beginning with the ones digit.** Let's look at a simple example to get the first step down:

Example: Multiply 37 x 6 = ?

First put this into column form:

$$\begin{array}{r} 37 \\ \times\,6 \\ \hline ? \end{array}$$

You will multiply each digit in the top number (the multiplicand) starting with the ones place by the bottom number (the multiplier).

We can equivalently write this as:

$$\begin{array}{l} 3 \text{ tens} + 7 \text{ ones} \\ \times \qquad\quad 6 \text{ ones} \\ \hline \qquad\qquad ? \end{array}$$

$$\begin{array}{l} 3 \text{ tens} + 7 \text{ ones} \\ \times \qquad\quad 6 \text{ ones} \\ \hline \qquad\quad 42 \text{ ones} \end{array}$$

Multiply the 7 ones by the 6 ones to obtain 42 ones.

$$\begin{array}{l} 3 \text{ tens} + 7 \text{ ones} \\ \times \qquad\quad 6 \text{ ones} \\ \hline 18 \text{ tens} + 42 \text{ ones} \end{array}$$

Multiply the 3 tens (30) by the 6 ones = 18 tens (180).

Combine 18 tens + 42 ones as 180 + 42 = 222. **So 37 x 6 = 222.**

Let's try this same example again using a slightly different approach which uses something like "**carrying**" from addition:

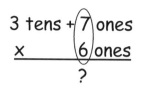

3 tens + 7 ones
x   6 ones
        ?

Multiply the 7 ones by the 6 ones to obtain 42 ones.

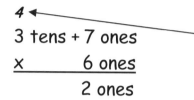

3 tens + 7 ones
x   6 ones
      2 ones

Instead of writing 42 ones, write 2 ones and **carry** 4 tens to the tens column.

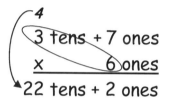

3 tens + 7 ones
x   6 ones
22 tens + 2 ones

Multiply 3 tens by 6 ones = 18 tens
Add carried 4 tens = 22 tens (220 ones).

Combine 22 tens + 2 ones as 220 + 2 = 222. **So 37 x 6 = 222.**

We can also see this multiplication example simply as:

  4
  37
x 6
  2

Multiply 7 x 6 = 42
   and **carry the 4**.

  4
  37
x 6
222

Multiply 3 x 6 = 18. Add carried 4 to get 22.
The 2 in 22 tens carries to hundreds column.
Therefore, 37 x 6 = 222.

Let's do one more simple example:

Example: 73 x 7 = ?

73   Put in column format.
x 7
___

2
7⟨3⟩   Multiply 3 x 7 = 21
x⟨7⟩   and **carry the 2**.
___
1

2
7̸3̸   Multiply 7 x 7 = 49. Add carried 2 to get 51.
x̸7̸   The 5 in 51 tens carries to hundreds column.
511   **Therefore, 73 x 7 = 511**.

What if there are 2 digits in the multiplier? Just add a step.

If there is more than one digit in the multiplier, we multiply by
each digit in the multiplier separately (beginning on the right) to
create **partial products**. Then, we **add the partial products**.
Each partial product must be **aligned** with the right end of the
multiplier digit so all numbers are in the correct columns.

Example: 46 x 38 = ?

We write this in column format and first multiply 46 x **8** using the carrying technique discussed above to get the first partial product. Next multiply 46 x **30** to get the second partial product. Then, add the two "partial products" to get the answer.

We have multiplied 46 by 8 to get 368. Next multiply 46 by 30. Note that 30 is 3 tens, so it looks like we are multiplying 46 by the 3, but the 3 is in the 10s place.

Notes on last step:

The 138 is really 1,380 (there is a placeholder 0 in the ones place). During the addition of partial products 368 + 1,380 we carried a 1 to the hundreds column when adding 6 tens + 8 tens = 14 tens, which was added to the hundreds: 3 + 3 + 1 = 7.

Therefore, 46 x 38 = 1,748

Example: 95 x 26 = ?

Write in column format. Multiply 95 x **6** to get the first partial product. (Remember to carry.) Next multiply 95 x **20** to get the second partial product. Then, add the two partial products to get the answer.

$$
\begin{array}{r}
95 \\
\times\ 26 \\
\hline
?
\end{array}
$$

First multiply
5 x 6 = 30
Carry the 3.

$$
\begin{array}{r}
\overset{3}{9}5 \\
\times\ 26 \\
\hline
0
\end{array}
$$

Then multiply
9 x 6 = 54
Add carried 3
to get 57.

$$
\begin{array}{r}
\overset{3}{9}5 \\
\times\ 26 \\
\hline
570
\end{array}
$$

We have multiplied 95 by 6 to get 570. Next multiply 95 by 20. Remember that 20 is 2 tens, so it looks like we are multiplying 95 by the 2, but the 2 is in the 10s place.

$$
\begin{array}{r}
95 \\
\times\ 26 \\
\hline
570
\end{array}
$$

Multiply 5 x 2
= 10. Put the 0
in the 10s
column below.
Carry the 1.

$$
\begin{array}{r}
\overset{1}{9}5 \\
\times\ 26 \\
\hline
570 \\
0
\end{array}
$$

Multiply 9 x 2 = 18
Add 1 = 19. Write 19
in 100s and 1,000s
columns. Add partial
products:
570 + 1,900 = 2,470

$$
\begin{array}{r}
\overset{1}{9}5 \\
\times\ 26 \\
\hline
570 \\
1900 \\
\hline
2,470
\end{array}
$$

Notes on last step:
The 190 is really 1,900 (there is a placeholder 0 in the ones place). During the addition of partial products 570 + 1,900 we carried a 1 when adding 5 + 9 = 14, which was added to the thousands: 1 + 1 = 2.

Therefore, **95 x 26 = 2,470**

What if there are 3 digits in the multiplier? Just add a step.

Example: Your city lot is 125 feet by 62 feet. How many square feet is it?

62 ft

125 ft

Let's set up our column format with the 125 in the multiplier. (Normally we would write the column format with the smaller number on the bottom, but we are learning 3-digit multipliers.)

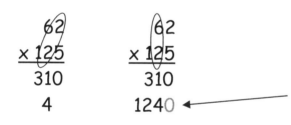

62	1	1
x 125	62	62
?	x 125	x 125
	0	310

Multiply 2 x 5 = 10. Carry 1 ten.
Multiply 6 x 5 = 30. Add 1 = 31.
First partial product is 310.
Clear the carried numbers.

62	62
x 125	x 125
310	310
4	1240

Multiply 2 x 2 = 4.
Multiply 6 x 2 = 12.
Second partial product is 1,240.
Note the 0 in 124 tens gives 1,240, since we multiplied 62 by 20.

62	62
x 125	x 125
310	310
124	1240
2	6200
	7750

Multiply 2 x 1 = 2. Multiply 6 x 1 = 6.
Third partial product is 6,200.
(Note: We multiplied 62 by 100.
The 0's are here even though we ignored them during multiplication.)
Now add partial products
310 + 1,240 + 6,200 = 7,750.
Therefore, **125 ft x 62 ft = 7,750 ft.**

What if one of the digits in the multiplier is a zero?

Just set up the multiplication using column format. Since
0 x 0 = 0, you can enter a row of zeros when you come to the
zero in the multiplier. An alternative is to skip the line, while
being careful to adjust the columns. Let's see:

**Example: You have 36 balls of yarn. Each has 108 yards. How
many yards total?**

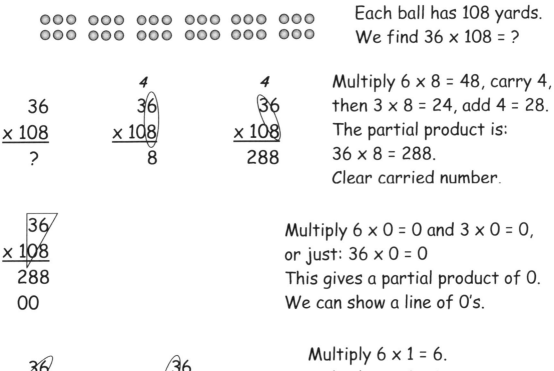

Each ball has 108 yards.
We find 36 x 108 = ?

Multiply 6 x 8 = 48, carry 4,
then 3 x 8 = 24, add 4 = 28.
The partial product is:
36 x 8 = 288.
Clear carried number.

Multiply 6 x 0 = 0 and 3 x 0 = 0,
or just: 36 x 0 = 0
This gives a partial product of 0.
We can show a line of 0's.

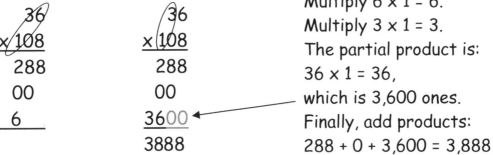

Multiply 6 x 1 = 6.
Multiply 3 x 1 = 3.
The partial product is:
36 x 1 = 36,
which is 3,600 ones.
Finally, add products:
288 + 0 + 3,600 = 3,888

Therefore, 36 x 108 = 3,888.

Note: An alternative to writing a line of 0s is to just insert a 0 under the 0 in the multiplier. Always remember to be careful that your columns are properly aligned.

Here is how the previous example looks with just inserting a zero rather than a whole line of zeros:

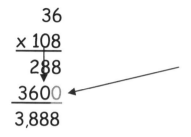

The first partial product is 36 x 8 = 288.
The second partial product is 36 x 0 = 0.
The 0 is placed under the 0 in the multiplier.
The third partial product is 36 x 1 = 36, but this 36 is really 3600 since it is 36 x 100.
So it gets aligned under the hundreds.

Let's look again at what happens if you multiply by a number ending in zero, such as 10 or 100?

In our base 10 system, each place to the left has a value 10 times higher.

So, a 4 in the 10s place is really 40, which is 10 times more than a 4 in the ones place.

Likewise a 4 in the 100s place means 400, which is 100 times a 4 in the ones place. Do you see a pattern here?

Look at the multiples of 1 in the table below. You can see how each number is 10-times greater as you move down and left:

	Ten Millions	Millions	Hundred Thousands	Ten Thousands	Thousands	Hundreds	Tens	Ones
1								1
1 x 10							1	0
1 x 100						1	0	0
1 x 1,000					1	0	0	0
1 x 10,000				1	0	0	0	0
1 x 100,000			1	0	0	0	0	0
1 x 1,000,000		1	0	0	0	0	0	0
1 x 10,000,000	1	0	0	0	0	0	0	0

We can practice multiplying numbers by 10, 100, 1,000, and 10,000. Let's look at the numbers 4, 77, and 987, and see what happens when we multiply them by 10, 100, 1,000, and 10,000.

To multiply a **whole number by 10**, just put **1 zero** on the right. This has the effect of moving the number to the left by one place on the above chart, making each digit **worth 10 times more**.

$$4 \times 10 = 40; \quad 77 \times 10 = 770; \quad 987 \times 10 = 9{,}870$$

To multiply a **whole number by 100**, put **2 zeros** on the right. This makes the number **worth 100 times more**.

$$4 \times 100 = 400; \quad 77 \times 100 = 7{,}700; \quad 987 \times 100 = 98{,}700$$

To multiply a **whole number by 1,000**, put **3 zeros** on the right. This makes the number **worth 1,000 times more**.

$$4 \times 1{,}000 = 4{,}000; \quad 77 \times 1{,}000 = 77{,}000; \quad 987 \times 1{,}000 = 987{,}000$$

To multiply a **whole number by 10,000**, put **4 zeros** on the right. This makes the number **worth 10,000 times more**.

$$4 \times 10{,}000 = 40{,}000; \quad 77 \times 10{,}000 = 770{,}000; \quad 987 \times 10{,}000 = 9{,}870{,}000$$

We see that when multiplying by multiples of 10, such as 1,000 or 10,000, we can put in the number of zeros on the right.

Example: Find 5×10 = ?, 24×100 = ?, and $357 \times 100,000$ = ?

$5 \times 1\underline{0}$ = $5\underline{0}$

$24 \times 1\underline{00}$ = $2,4\underline{00}$

$357 \times 1\underline{00,000}$ = $35,7\underline{00,000}$

Note that the number of zeros in the multiplier and the product are the same. See underlined 0's.

Centimeters, meters, and kilometers are units of measure used in science. There are about 3.28 feet per meter, about 3,280 feet per kilometer, 100 centimeters per meter, and 1,000 meters per kilometer. Let's do some examples using these commonly used units of measure:

Example: How many meters (m) are in 258 kilometers (km)?

Since 1 km = 1,000 m

then 258 km = 258,000 m

Using multiplication: 258 km x 1,000 m/km = 258,000 m

Example: How many centimeters (cm) in 43 meters?

Since 1 m = 100 cm

then 43 m = 4,300 cm

Using multiplication: 43 m x 100 cm/m = 4,300 cm

Multiplying 3 Numbers

We can visualize multiplying three numbers: 5 x 3 x 4 = ?

Imagine a garden that is **5 feet by 3 feet**, and each **square foot contains 4 flowers**. How many flowers are there? Let's draw it.

	1 4	2 8	3 12	4 16	5 20
3	6 24	7 28	8 32	9 36	10 40
	11 44	12 48	13 52	14 56	15 60

5

The <u>top number</u> in each square shows how to count the number of squares. The <u>bottom number</u> counts the number of flowers, counting up by fours. You can see that there are 15 square feet and 60 flowers in your garden.

How do you figure out 5 x 3 x 4 without counting? It's a bit like adding three numbers. **Multiply the first two numbers, then multiply the answer by the third number. So:**

to find **5 x 3 x 4**, first do **5 x 3 = 15**, then do **15 x 4 = 60**

Note: 5 x 3 gives you the number of square feet (15). Then you multiply 15 square feet by the number of flowers per square foot (4) to get 60 flowers!

Note: Because multiplication is like addition (but unlike subtraction), it can be done in any order. That means:

$5 \times 3 \times 4 = 5 \times 4 \times 3 = 4 \times 3 \times 5 = 4 \times 5 \times 3 = 3 \times 5 \times 4 = 3 \times 4 \times 5$

Example: How many 1-inch-square blocks does it take to build a tower if its base is 6 blocks by 5 blocks, and its height is 7 blocks?

The Tower

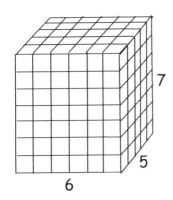

To find the total blocks, first find the number of blocks in the base (which is the number of blocks in each horizontal layer):

6 x 5 = 30 blocks per layer

Next multiply the blocks per layer by the number of layers, which is the height:

30 blocks/layer x 7 layers = 210 blocks

Therefore, 6 x 5 x 7 = 210. There are 210 blocks.

Multiplying 3 numbers also describes a volume.

Multiplying 3 numbers is used to find the volume of a solid object with rectangular sides:

Volume = length x width x height

Example: What is the volume of a box with a 3 by 4 meter base and a height of 5 meters.

To find volume we calculate length x width x height = 3 x 4 x 5
To find 3 x 4 x 5, first do 3 x 4 = 12, then do 12 x 5 = 60
Therefore, 3 x 4 x 5 = 60.

Multiplying 4 or More Numbers

When multiplying 4 or more numbers together, you can multiply pairs (in any order) until each number has been multiplied.

Example: In the previous example about the 6 by 5 by 7 tower, if each block in the tower weighs 8 grams, what does the tower weigh?

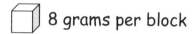 8 grams per block

This involves multiplying the 3 tower dimensions by the weight of each block. This means we have 4 numbers to multiply:

$$6 \times 5 \times 7 \times 8 = ?$$

In the previous example we learned that:

$$6 \times 5 \times 7 = 210$$

We found this by first multiplying the horizontal blocks/layer:

$$6 \times 5 = 30 \text{ blocks per layer}$$

Then multiplying the result with the vertical dimension:

$$30 \text{ blocks/layer} \times 7 \text{ layers} = 210 \text{ blocks}$$

We now multiply the 210 blocks by the weight per block:

$$210 \text{ blocks} \times 8 \text{ grams/block} = 1,680 \text{ grams}$$

If we show the multiplication all together we have:

$$6 \times 5 \times 7 \times 8 = 1,680 \text{ grams}$$

Example: Each month for 12 years, 12 airplanes arrive in Bushtown carrying 12 tourists each. If each passenger spends $12 before leaving, how much money is spent during those 12 years by the tourists?

If we multiply these all together we will get the dollars spent by all tourists. Note that you can keep track of the units to see that they cancel each other. The units are:

years x months-per-year x planes-per-month x tourists-per-plane
x $-per-tourist = total $

Which we can equivalently write as:

years x months/year x planes/month x tourists/plane x $/tourist = $

Now put in the numbers:

(12 years) x (12 months per year) x (12 planes per month) x
(12 tourists per plane) x (12 $ per tourist) = ?

Which we can equivalently write as:

(12 yrs)(12 months/yr)(12 planes/month)(12 tourists/plane)(12 $/tourist)

= 12 x 12 x 12 x 12 x 12 = ?

Multiply the 1st and 2nd 12s: 12 x 12 = 144
Then multiply the product by the 3rd 12: 144 x 12 = 1,728
Then multiply the product by the 4th 12: 1,728 x 12 = 20,736
Finally multiply the product by the 5th 12: 20,736 x 12 = 248,832
So the tourists spent a total of $248,832.

Extra Credit:

We have learned that when a number is added to itself many times, we can use multiplication: $12 + 12 + 12 + 12 + 12 = 12 \times 5$

Similarly, when a number is multiplied by itself many times, we can show it by using what is called an **exponent**.

Instead of writing $12 \times 12 \times 12 \times 12 \times 12$ you can just write 12^5. Superscript "5" tells how many 12s are multiplied together.

Example: $4^{10} = 4 \times 4 \times 4 \times 4 \times 4 \times 4 \times 4 \times 4 \times 4 \times 4$
Where 4^{10} represents ten 4s multiplied together.

Example: How would you write 1 trillion?
1 trillion = 1,000,000,000
$= 10 \times 10 \times 10 \times 10 \times 10 \times 10 \times 10 \times 10 \times 10 = 10^9$
Where 10^9 represents 9 tens multiplied together.

Multiplying Even and Odd Numbers

We discussed even and odd numbers in Sections 1.3 and 2.5. What about multiplying even and odd numbers? Let's look! If E means even and O means odd, show the combinations:

$E \times E = E$ $E \times O = E$ $O \times E = E$ $O \times O = O$

Why should all the products be even except odd x odd, O x O? Multiplication is adding over and over. When even numbers are added together any number of times, there is never an "odd man out". That means $E \times E = E$. But what about $E \times O = E$, $O \times E = E$, and $O \times O = O$. Can you think of a way to prove that $E \times O = E$, $O \times E = E$, and $O \times O = O$? What about the swimming buddies?

Draw an odd number, 3, groups of swimmers. Give each group, an even number, 4, swimmers. This is: 3 x 4. Is it even?

Each group of 4 swimmers has 2 pairs of buddies. With the 3 groups all together, everyone has a buddy. There are no unpaired swimmers.

We see 3 x 4 = 12 buddies. All swimmers have a buddy. There are with no unpaired buddies. This shows it's even.

Is the same true if you have 4 groups of 3 swimmers each? Yes:

Extras pair off in group 1 with 2, and group 3 with 4. This leaves no unpaired swimmers.

See 12 buddies can still each pair up, showing it's still even.

Now let's look at an odd number of odd-sized groups.
Look at 3 x 5:

Here everyone has a buddy except the "odd" man in the 5th group. So you have an odd number. In fact, 5 x 3 = 15, which is an odd number.

3.2. Divide

What is Division?

Do you know what the symbols for "divide" look like? They are:

$\overline{)}$, ÷, and /

Can you show 63 divided by 9 using the different symbols?

$9\overline{)63}^{\,?}$, $\frac{63}{9}$, 63 ÷ 9, or 63 / 9.   Are these all equivalent? Yes.

Let's try to understand what division means. You may have been taught that **division is the "opposite" of multiplication.**

This is true. If we multiply: 9 x 7 = 63

Reversing the operation, we get: 63 ÷ 9 = 7 and 63 ÷ 7 = 9

What does division do? **It calculates how many times one number is present, or "fits", in another number.** Let's look:

 How many groups of 3 dots are present in 9 dots?
We can see there are 3 groups of 3 dots in 9 dots!
9 ÷ 3 = 3   or   9/3 = 3   also   3 x 3 = 9

Terminology: Every division problem has 2 numbers and also an answer. Using the numbers 63 ÷ 9 = 7, the technical names are:

63 ("**dividend**") ÷ 9 ("**divisor**") = 7 ("**quotient**")

$$\text{divisor}\overline{)\,\text{dividend}}^{\,\text{quotient}} \qquad \frac{\text{dividend}}{\text{divisor}} = \text{quotient}$$

Can you think of a way to visualize division? What about looking at how many groups of 3 or 4 dots are in 12 dots?

$$12 \div 4 = 3$$

$$12 \div 3 = 4$$

4 dots fit into 12 dots 3 times. See 3 groups of 4 dots in 12 dots.

3 dots fit into 12 dots 4 times. See 4 groups of 3 dots in 12 dots.

Let's see division by going into the kitchen and making omelets!

Example: You have 12 eggs. How many 3-egg omelets can you make?

To do this, you divide the eggs into groups of 3 eggs each. You can write 12 eggs ÷ 3 eggs, or 12/3 = ? You can draw it out as:

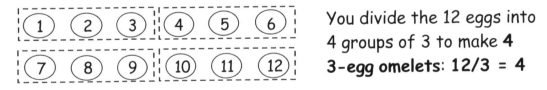

You divide the 12 eggs into 4 groups of 3 to make **4 3-egg omelets: 12/3 = 4**

Therefore, you can make four 3-egg omelets.

Example: Suppose you want smaller omelets. How many 2-egg omelets can you make from the 12 eggs?

You can divide the 12 eggs into groups of 2 eggs each: 12/2 = ?

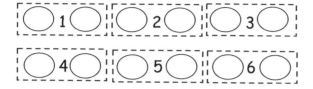

You divide the 12 eggs into 6 groups of 2 to make **6** **2-egg omelets: 12/2 = 6**

Therefore, you can make six 2-egg omelets.

Example: You have lumberjacks over who want 4-egg omelets. With 12 eggs how many omelets can you serve?

With your dozen, or 12, eggs you divide by 4: 12/4 = ?

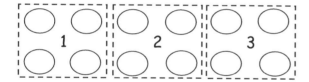

You can divide 12 eggs into 3 groups of 4 to make **3** **4-egg omelets: 12/4 = 3**

Therefore, you can make three 4-egg omelets.

Are there other ways to divide your 12 eggs? Yes, you could also divide the 12 eggs into 2 groups of 6 to make two 6-egg omelets, written 12/6 = 2. You could even make 1 huge 12-egg omelet, written 12/12 = 1! Also, with your 12 eggs you could make twelve 1-egg mini-omelets, written 12 ÷ 1 = 12 mini-omelets.

As you can see, dividing is like separating a larger number into many small groups of <u>equal size</u>. In other words, when dividing something into smaller groups or objects, the smaller groups or objects <u>must</u> be the same size. They cannot be different sizes.

Example: There are 35 kids at a basketball camp. How many teams of 5 kids each can you form?

We are asking 35 / 5 = ? Let's count them from left to right:

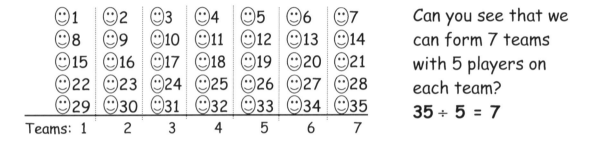

Can you see that we can form 7 teams with 5 players on each team?

35 ÷ 5 = 7

Therefore, you can form 7 teams with 5 players each.

Here's a simple example from Chapter 1 to help you see division.

Example: How many groups of 2 faces are there in 6 faces?

6 faces

2 faces ▯ go into ⟶ ▯ ▯ ▯ ⟶ 3 times

You can see 3 groups of 2 faces in 6 faces.
2 faces go into 6 faces 3 times.
This shows there are three 2s in 6.

Example: How many ounces (oz) are in each piece of pizza if you cut a 32 oz pizza into 8 equal pieces?

We are asking: 32 oz ÷ 8 pieces = ? oz per piece

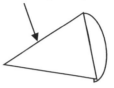

Pizza

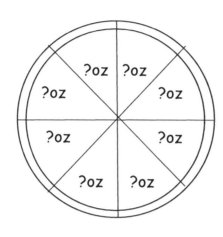

We can write the 8 pieces as:

?oz + ?oz + ?oz + ?oz + ?oz + ?oz + ?oz + ?oz = 32 oz

By trying different numbers for the ounces we can find 4oz eight times equals 32oz.

So: 4oz + 4oz + 4oz + 4oz + 4oz + 4oz + 4oz + 4oz = 32 oz

To calculate oz/piece divide total ounces by number of pieces:

32 oz ÷ 8 pieces = 4 oz/piece

Dividing a Number By Itself Always Equals 1

Example: 100/100 = ?

This is like asking, "How many $100 bills are in a $100 bill?"

Exactly one $100 bill will "fit" into a $100 bill.

A number fits into itself exactly 1 time!

What About Dividing a Number By 1?

Any Number Divided By 1 Is Always That Number!

Example: If you have 100 stars, how many people can you give 1 star to?

1 goes into 100
exactly 100 times.

$$100 \div 1 = 100$$

Therefore, you can give 1 star each to 100 people.

Example: If you have 100 $1 bills, how many people can you give $1 to?

$$100 \div 1 = 100$$

Therefore, your 100 $1 bills can be given to 100 people.

Example: Divide $25,867 \div 1 = ?$

$$25,867 \div 1 = 25,867$$

We see that dividing a number by 1 is always that number.

Can A Number Line Show How Division Works? Yes!

Let's look at 15/3 = ?

The division, 15/3 = ?, asks, "How many 3s are there in 15?"

To find out, you can start with 15 and subtract 3 over and over until you get to zero. Let's show that on the number line:

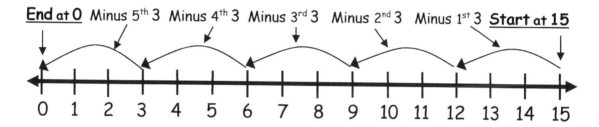

After you have subtracted **3 five times**, you get to zero.
We see 3 fits into 15 **five** times. This means: **15/3 = 5**.

Let's also show 15/5 = ?
You begin again at 15 and subtract 5 over and over until you get to zero.

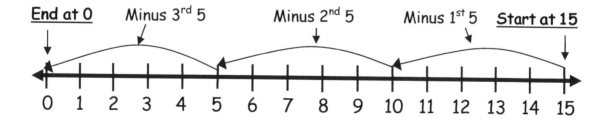

After subtracting **5 three times**, you get to zero.
We see 5 fits into 15 **three** times. This means: **15/5 = 3**.

Remainders

What happens when you divide one number into a second number,
but it won't go in exactly and there is extra left over?
The left over is called a "**remainder**". Let's look:

14 ÷ 3 = 4 and a remainder 2

See 14 divided by 3 gives
4 groups and 2 remaining.

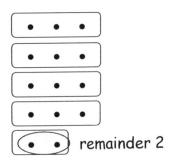

remainder 2

We can also see this as:
14 divided into 3's results in
4 groups with 2 left over.

14 ÷ 4 = 3 and a remainder 2

See 14 divided by 4 gives
3 groups and 2 remaining.

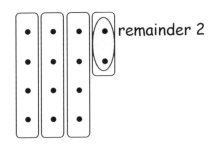

remainder 2

We can also see this as:
14 divided into 4's results in
3 groups with 2 left over.

14 ÷ 7 = 2 and a remainder 0

See 14 divided by 7 gives 2 groups of 7 and 0 remaining.

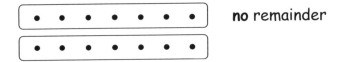

no remainder

We can also see this as:
14 divided into 7's results in 2 groups with nothing left over.

Example: Show 24 ÷ 6 = 4, 24 ÷ 4 = 6, and 27 ÷ 4 = 6 r3

Let's look at the division of 24 ÷ 6 = 4 and 24 ÷ 4 = 6
as dividing 24 stars into either:

4 groups of 6 stars   or   6 groups of 4 stars

6 goes into 24 **four** times and 4 goes into 24 **six** times

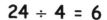

24 ÷ 6 = 4   **24 ÷ 4 = 6**

24 ÷ 6
divides 24
stars into
4 groups of
6 stars each

24 ÷ 4
divides 24
stars into
6 groups of
4 stars each

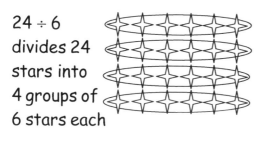
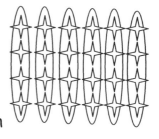

We see that 24 divides exactly by 4 or 6.

What about 27 ÷ 4 = 6 r3 It has 3 left over as a **remainder**.
We can show 27 ÷ 4 = 6 with 3 left over using stars:

27 ÷ 4 = 6 with 3 left over

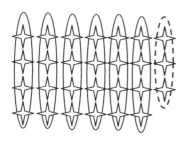

27 ÷ 4 yields 6 groups of 4 stars each,
but there are 3 stars left. The extra 3
are too few to make another group of 4.
The extra stars are called the "remainder".

4 goes into 27 **six** times with 3 left over

Example: You have $15 and want to buy $2 gifts for friends. How many friends can you buy $2 gifts for?

If you count 15 $1 bills and group them into 2s, what do you see?

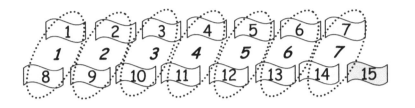

You can buy 7 gifts and you also have **$1 left over** (as a remainder).

See that 15 divided by 2 is *not just* 7 since there is 1 left over.
Therefore, you can buy 7 gifts, but will have $1 left over.

So, what do we do with a remainder and what does it mean?

We saw that when we divide a number into groups of a particular size, it may not divide evenly or perfectly. After most of the number is divided into groups, there may be a **remainder** left over. The remainder is always smaller than one of the groups. (If it was larger than the size of a group, it could be used to form another group.) You can think of a remainder as a portion or fraction of a whole group!

How many ways can you write a remainder? Two.
The first is to just **show or write down the remainder**.
The second is to **show the remainder as equal to some part or fraction of a group**. (Chapter 4 explains fractions in detail.)

Example: Earlier we drew 14÷3 and 14÷4. Write the remainders as a part, or fraction, of a group.

Again, we see:

14/3 = 4 and remainder 2 dots,   14/4 = 3 and remainder 2 dots

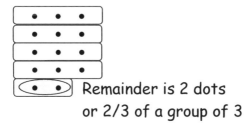

Remainder is 2 dots
or 2/3 of a group of 3

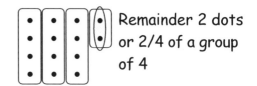

Remainder 2 dots
or 2/4 of a group
of 4

Write remainders as part of a group, or: part/whole group

14/3 = 4 and 2/3 of a group, or just 14/3 = 4 2/3

14/4 = 3 and 2/4 of a group, or just 14/4 = 3 2/4

Example: Write and show 15/2 with a remainder and write the remainder as a part or fraction of a group.

Written with its remainder: **15/2 = 7 and remainder 1**

★★★★★★★★★ ←—— The remainder is 1,
★★★★★★★★   or 1/2 of a group.

Writing its **remainder as a part or fraction of a group**:

15/2 = 7 and 1/2 of a group or just **15/2 = 7 1/2**

Example: You have 19 pizzas and need to divide them equally among 6 friends. How much pizza will each friend get?

This is: 19 pizzas / 6 friends = ? pizzas/friend

You know: 6 x 3 = 18. So: 18/6 = 3 or 6 goes into 18 three times.

Since you have 19 pizzas, there's 1 pizza left as a **remainder** to divide among the 6 friends. We write this as:

19 pizzas / 6 friends = 3 whole pizzas plus 1/6 pizza for each friend.
Remember, 6 sixths = 6/6 = 1. **Each friend gets 3 1/6 pizzas.**

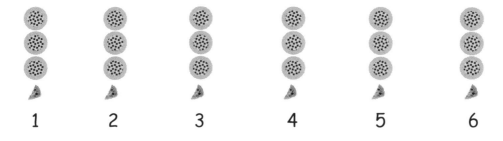

1   2   3   4   5   6

Example: What is 15 ÷ 20 = ?

We cannot fit 20 into 15

This means 15/20 = 0 with a remainder of 15

But the remainder 15 is really 15/20

This means 15 ÷ 20 is the same as 15/20

This shows us that a division problem is a fraction

and that a fraction is a division problem!

Note: 15/20 can be "reduced" to 3/4 or divided and put in decimal form as 0.75, which we will learn about in Chapter 6.

How Do You "Check" Your Answer In Division?

To check if your answer (quotient) is correct, multiply it by the bottom number (divisor). You should get the top number (dividend).

$$\frac{24 \text{ (dividend)}}{8 \text{ (divisor)}} = 3 \text{ (quotient)}$$

Example: You think 24/8 = 3, but how can you be certain?

To see if $\dfrac{24}{8} = 3$ multiply 3 x 8. Does 3 x 8 = 24? Yes!

Example: Does 72 ÷ 9 = 8 ?

To see if $\dfrac{72}{9} = 8$ multiply 8 x 9. Does 8 x 9 = 72? Yes!

Example: Does $1.00 ÷ 4 = 25 cents?

Note that dollars are in dollar $ units and cents are in cents units. We know there are 100 cents in 1 dollar, or $1.00.
To be sure what we divide makes sense, we need to divide numbers in the same units. Rewrite the example all in cents:

Does 100 cents ÷ 4 = 25 cents?

To see if $\dfrac{100 \text{ cents}}{4} = 25$ cents multiply 25 cents x 4 = ?

Yes! 25 cents x 4 = 100 cents = $1.00. We know 4 quarters make a dollar! **Yes, $1.00 ÷ 4 = 25 cents.**

Zero and Division (zero is weird because it shows nothing)

What happens if you divide by Zero?

Dividing by zero (nothing) is "undefined"

number ÷ 0 = "undefined"   or $\dfrac{\text{number}}{0}$ = "undefined"

2/0 = undefined   357/0 = undefined   8,888,888/0 = undefined

Why is dividing by zero "undefined"? Think about: $12 \div 0 = ?$

$12 \div 0$ says "How many zeros are there in 12?" What does that really mean? If you have a 12-pound block of cheese and slice it infinitely thin so each slice weighed 0 pounds, how may slices would you have?

Each slice is infinitely thin and weighs 0 pounds

You may be tempted to say "an infinite number" so that
12 pounds ÷ 0 = ∞. (The symbol for **infinity** is "∞".)

What if you check that answer? Does ∞ x 0 = 12 ? No.

It is silly. If it were true, it would mean that when dividing any number by zero the answer would always be infinity. Can you really say:

$12 \div 0 = \infty$, $6 \div 0 = \infty$, $60 \div 0 = \infty$,   **No**

These are <u>not</u> all equal to infinity. Further, if we check by multiplying, does:

∞ x 0 = 12,   ∞ x 0 = 6,   ∞ x 0 = 60 ?   **No**

0 x ∞ does not equal any number!

We see that dividing a number by 0 cannot equal ∞, because when we reverse the division to check our answer, 0 x ∞ does **not** equal any number.

Therefore, mathematicians have decided that **dividing by 0 is "undefined"** because technically it doesn't make sense! When performing operations on equations (as in algebra or calculus), if you divide by something that equals zero, the resulting equations will not be valid. Dividing a number by zero is therefore called "undefined":

$$number ÷ 0 = undefined$$

What about zero divided by a number?

Zero (nothing) divided by a number is zero

$$0 ÷ number = 0 \quad\quad or \quad\quad \frac{0}{number} = 0$$

$0/2 = 0 \quad\quad 0/20 = 0 \quad\quad 0/357 = 0 \quad\quad 0/8,888,888 = 0$

Why is **0 ÷ number = 0**?
Because nothing divided by a certain number is still nothing.
Let's think about this. If you have zero basketball players, how many teams of 5 players can you form?
We would write this as: 0 ÷ 5 = ?
Of course, you cannot form any teams if you have no players.
Therefore, 0 ÷ 5 = 0.

Simple Division Calculations You Can Picture

As we saw earlier, a **number line** can be used to map out and see division of very small numbers. (Of course, calculating division of larger numbers using a number line is not practical.)

A **Multiplication Table** is also a way to **see** division for small numbers **by working backwards!**

Multiplication Table

×	0	1	2	3	4	5	6	7	8	9
0	0	0	0	0	0	0	0	0	0	0
1	0	1	2	3	4	5	6	7	8	9
2	0	2	4	6	8	10	12	14	16	18
3	0	3	6	9	12	15	18	21	24	27
4	0	4	8	12	16	20	24	28	32	36
5	0	5	10	15	20	25	30	35	40	45
6	0	6	12	18	24	30	36	42	48	54
7	0	7	14	21	28	35	42	49	56	63
8	0	8	16	24	32	40	48	56	64	72
9	0	9	18	27	36	45	54	63	72	81

In the Table we can see that 4 x 6 = 24 and also 6 x 4 = 24.
We can go from 24 ÷ 6 to get to 4, which is **24 ÷ 6 = 4**.
We can also go from 24 ÷ 4 to get to 6, which is **24 ÷ 4 = 6**.

Let's look at some examples using the Multiplication Table.

Multiplication Table

×	0	1	2	3	4	5	6	7	8	9
0	0	0	0	0	0	0	0	0	0	0
1	0	1	2	3	4	5	6	7	8	9
2	0	2	4	6	8	10	12	14	16	18
3	0	3	6	9	12	15	18	21	24	27
4	0	4	8	12	16	20	24	28	32	36
5	0	5	10	15	20	25	30	35	40	45
6	0	6	12	18	24	30	36	42	48	54
7	0	7	14	21	28	35	42	49	56	63
8	0	8	16	24	32	40	48	56	64	72
9	0	9	18	27	36	45	54	63	72	81

Example: What is 73 ÷ 8 = ?

Looking at the Multiplication Table, you can see that:
8 x 9 = 72 or **72/8 = 9**. (See ovals above.)
Since **73 is 1 more than 72**, then **73/8 = 9 plus remainder 1**.
We can just write: **73/8 = 9 1/8**.

Example: What is 78 ÷ 8 = ?

Using our Multiplication Table, you can see that:
8 x 9 = 72 or **72/8 = 9**. (See ovals above.)
Since **78 is 6 more than 72**, then **78/8 = 9 plus remainder 6**.
We can just write: **78/8 = 9 6/8**.

Multiplication Table

×	0	1	2	3	4	5	6	7	8	9
0	0	0	0	0	0	0	0	0	0	0
1	0	1	2	3	4	5	6	7	8	9
2	0	2	4	6	8	10	12	14	16	18
3	0	3	6	9	12	15	18	21	24	27
4	0	4	8	12	16	20	24	28	32	36
5	0	5	10	15	20	25	30	35	40	45
6	0	6	12	18	24	30	36	42	48	54
7	0	7	14	21	28	35	42	49	56	63
8	0	8	16	24	32	40	48	56	64	72
9	0	9	18	27	36	45	54	63	72	81

Example: How many boxes does it take to hold 70 bricks if each box holds 8 bricks?

In the Multiplication Table see that: **8 x 8 = 64 or 64/8 = 8**.
(See solid ovals above.)

Since **70 is 6 more than 64**, then **70/8 = 8 plus remainder 6**.
To get all the bricks into boxes, you will need not just 8 boxes but actually 9 boxes, with the 9th box holding the extra 6 bricks.

Example: How many 3-egg omelets can you make with 23 eggs?

In the previous Table (dashed ovals) see:
3 x 7 = 21 or 21/3 = 7.
Since **23 is 2 more than 21**, then **23/3 = 7 plus remainder 2.**

We can also draw this example using omelet pans.
Let's picture how many 3-egg omelets we can make with 23 eggs by placing 3 eggs in each pan:

| 1 2 3 | 4 5 6 | 7 8 9 | 10 11 12 | 13 14 15 | 16 17 18 | 19 20 21 | 22 23 | The 2 extra eggs are the remainder. |

You can count out 7 pans of 3 eggs each until you run out of 3-egg groups. Counting out the eggs shows we can make **7 three-egg omelets.**

We see that 3 divides into 21 exactly:

$$21 \text{ eggs}/3 \text{ eggs per pan} = 7 \text{ omelets}$$

Since we have 23 eggs and 23 - 21 = 2, then we have:

23/3 = 7 omelets and 2 extra eggs.

Long Division - A Reliable Process to Calculate Division

What do you do when you don't have or can't use a calculator, and you need to divide two numbers? **Long division** will let you divide any two numbers, even when you get remainders. All you need is a pencil and paper.

Written in long division, 16 ÷ 3 = 5 plus remainder 1 looks like:

```
Quotient ⟶   5
Divisor ⟶ 3)16
   Dividend   15
Remainder ⟶  1
```

Long division of larger numbers is generally done in steps:
 -divide the divisor into the leftmost part of the dividend it
   will go into, and put the quotient above;
 -multiply that quotient by the divisor, and put the product below;
 -subtract that product from dividend numbers above it; and
 -bring down the next number in dividend and divide by the divisor.

This process continues until all digits in the dividend are used.

In other words, when we do long division, we start by dividing the divisor into the leftmost part of the dividend number. Then move right as you divide multiply and subtract as described. Let's do some examples.

Example: Divide 168 ÷ 3 = ?

$$\begin{array}{r} ? \\ 3\overline{)168} \end{array}$$

Divide 3 into the leftmost part of 168 it will go into. 3 won't go into 1, but will go into 16.

3 goes into 16 about 5 times, but there is a remainder. Put a 5 above the 6 to show that 3 goes into the 16 five whole times.

To find the remainder, multiply quotient 5 by the divisor 3 (then we will subtract). But first: 5 x 3 = 15. Write the product 15 below the 16.

$$\begin{array}{r} 5 \\ 3\overline{)168} \\ \underline{15} \\ 1 \end{array}$$

Subtract the 15 from 16 above it. 16 - 15 = 1 remainder.

$$\begin{array}{r} 5 \\ 3\overline{)168} \\ 15\downarrow \\ 18 \end{array}$$

Bring down the next dividend number 8.

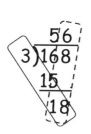

Divide the resulting 18 by divisor 3. 3 goes into 18 **exactly** 6 times: 18/3 = 6 Put the 6 over the 8 in the dividend. This gives 56 as the quotient of 168/3. All digits in the dividend have been used. **There is no remainder**.

Therefore, **168 ÷ 3 = 56** or **168/3 = 56**.
Check answer using multiplication: Does 56 x 3 = 168? Yes.

Example: If you have $369 to split equally between 3 charities, how much does each charity receive?

We just divide 369 ÷ 3. To use long division we put the dividend inside the bracket and the divisor to the left. The quotient will be put on top as it is computed.

 Divide 3 into the leftmost number in the dividend it will go into. 3 goes into 3 exactly 1 time: 3/3 = 1, no remainder. Enter 1 above 3 in the 100s place.

There is no remainder for 3/3 = 1 so we move on to the next dividend number. (If there was a remainder we would multiply the first quotient by the divisor, subtract the product, and then bring down the next number 6 in the dividend to divide by divisor 3. But we would end up with 3 - 3 = 0, and then would bring down the 6 to divide into.) Instead we just proceed with dividing the divisor 3 into 6.

 How many times does divisor 3 go into 6? 3 goes into 6 exactly 2 times: 6/3 = 2, no remainder. Put quotient 2 above the 6 in the 10s place.

Again, there is no remainder for 6/3 = 2, so we don't need to multiply, subtract, and bring down the next number 9 to divide by 3. Instead we can just proceed with dividing the divisor 3 into 9.

 How many times does divisor 3 go into 9? 3 goes into 9 exactly 3 times: 9/3 = 3, no remainder. Put quotient 3 above 9 in the 1s place. All digits in the dividend have been divided into and there were no remainders!

Therefore, 369 ÷ 3 = 123.

Let's check using multiplication. Does 123 x 3 = 369? Yes.

So each of your **3 charities can receive exactly $123**.

Let's do an example with remainders. In each division step, the remainder becomes part of the next step.

Example: Compute 478/3 using long division.

Divide divisor 3 into the leftmost part of dividend number. 3 goes into 4 one time: 4/3 = 1, with a **remainder**. Put the 1 above 4 in the 100s place.

To find the **remainder**, first multiply the 1 by divisor 3: 3 x 1 = 3. Put product 3 under the 4 in the hundreds place.

Subtract product 3 from 4 above it, resulting in the **remainder of 1**: 4 - 3 = 1. Bring down the next part of the dividend, 7 tens.

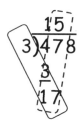

Divide the resulting 17 by divisor 3: 17/3 = 5, with a **remainder**. Put the 5 above 7 in the 10s place.

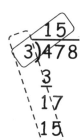

To find the **remainder**, first multiply the 5 tens by divisor 3: 5 x 3 = 15. Put product 15 under the 17.

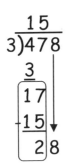

Subtract product 15 from 17 above it, resulting in the **remainder of 2**: 17 - 15 = 2.

Bring down the next part of the dividend, 8 ones.

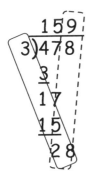

Divide the resulting 28 by divisor 3.

28/3 = 9, with a **remainder**.

Put the 9 over the 8 in the ones place.

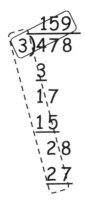

To find the **remainder**, first multiply the 9 ones by divisor 3: 9 x 3 = 27.

Put product 27 under the 28.

```
  159 r1
3)478
  3
  17
  15
  28
 -27
   1
```

Subtract product 27 from 28 above it giving the final **remainder of 1**: 28 - 27 = 1. Place remainder 1, r1, at the top next to the quotient. All digits in the dividend have been divided into.

Therefore, 478/3 = 159 r1 or 159 1/3

You're doing great! Let's try another LONG division.

Example: You have 2,368 beads to make 8 identical necklaces. What is the largest number of beads each necklace can have?

We can divide the total number of beads by the number necklaces we want to determine beads per necklace:

2,368 beads / 8 necklaces = ? beads/necklace

Note that the divisor 8 is larger than the first number 2 in the dividend, so we begin this division problem by dividing 8 into 23.

Divide divisor 8 into the leftmost part of dividend it goes into. 8 goes into 23 about 2 times: 23/8 = 2, with a **remainder**. Enter 2 above 3 in the 100s place.

Find the **remainder** by first multiplying the 2 by divisor 8: 2 x 8 = 16. Put the 16 under 23.

Subtract the 16 from 23 above it, resulting in the **remainder of 7:** 23 - 16 = 7.

Bring down the next part of the dividend, 6 tens.

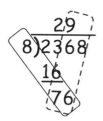

Divide the resulting 76 by divisor 8:
76/8 = 9 with a **remainder**.
Put the 9 over the 6 in the tens place.

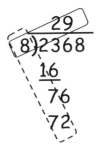

Find the **remainder** by multiplying the 9 by divisor 8: 9 x 8 = 72. Put the 72 under the 76.

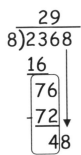

Subtract the 72 from 76 above it resulting in the **remainder of 4**: 76 - 72 = 4.

Bring down the next part of the dividend, 8 ones.

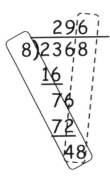

Divide the resulting 48 by divisor 8.
48/8 = 6, with **no remainder**.
Put the 6 over the 8 in the ones place.

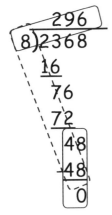

We can end here since there is **no remainder**.
But we can show this by multiplying the 6 ones by divisor 8: 6 x 8 = 48. Put product 48 under the 48. Then subtract 48 - 48 = 0
All digits in the dividend have been divided into, and there is no remainder.

Therefore, 2,368 beads / 8 necklaces = 296 beads/necklace.
Let's check: Does 296 x 8 = 2,368? Yes!

Estimates and Shortcuts

Sometimes you can estimate or take shortcuts when dividing.

Example: Divide 8,024 by 4 using a shortcut.

Let's dissect our number: 8,024 = 8,000 + 24.
This means we can divide 8,024/4 as:

$$8,000/4 + 24/4 = 2,000 + 6 = 2,006$$

Therefore, we can see that 8,024/4 = 2,006.

But we can also show 8,024/4 = 2,006 using long division:

Divide divisor 4 into leftmost part of dividend.
4 goes into 8 exactly 2 times: 8/4 = 2, **no
remainder**. Put 2 over 8 in the thousands place.

Multiply the 2 by divisor 4: 2 x 4 = 8
Put product 8 under the 8.

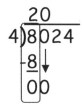

Subtract product 8 from 8 above it to show
remainder 0: 8 - 8 = 0.
Bring down the next dividend number, 0.

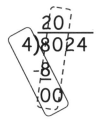

Divide divisor 4 into 00: 0/4 = 0.
Put the 0 over 0 in the 100s place.

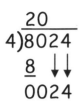

You can multiply 0 x 4 = 0, put 0 below 00, and then subtract 0 - 0 = 0. But it's quicker to stay on the same line and bring down the next dividend number 2. Since divisor 4 does not go into 2, put 0 above 2 in the quotient and bring down dividend number 4.

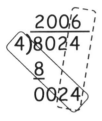

Divide the resulting 24 by divisor 4.
4 goes into 24 exactly 6 times: 24/4 = 6, **no remainder**. Put 6 over 24 in the ones place.

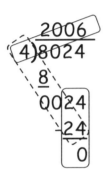

We can stop here since there is **no remainder**. But for practice, multiply the 6 ones by divisor 4: 6 x 4 = 24. Put 24 under the 24.
Then subtract 24 - 24 = 0.
All digits in the dividend have been divided into and there is no remainder.

Therefore, we show using long division: 8,024/4 = 2,006.
Let's check: Does 2,006 x 4 = 8,024? Yes!

Long Division with Large Divisors

When the divisor is larger than a single-digit number, we can still use the same long-division process involving smaller divisors, multiplying, subtracting, and bringing down dividend numbers. Let's practice:

Example: If you want to earn $9,876 in a year, what must you earn on average each month?

You divide total dollars you earn in a year by months per year to get dollars you must earn per month:

9,876 total dollars / 12 months = ? dollars/month

Let's do the long division:

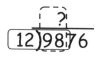

You start by dividing divisor 12 into the leftmost part of the dividend it goes into. You cannot divide 12 into 9, so you divide 12 into 98.

You may not remember how many 12s can "fit" into 98. You could look at an extended multiplication/division table and find the "times 12" line to see the largest number that is less than 98. Or you could use the **guess and multiply strategy**. Let's try that:

Suppose you guess 9. Well, 12 x 9 = 108. This is greater than 98, so it can't be right. You try 8. You remember 12 x 8 = 96. This is less than 98. How much less? 98 - 96 = 2. We'll use that:

12 goes into 98 about 8 times: 98/12 = 8 with a **remainder**. Put the 8 over the 8 in 98.

To find the **remainder**, multiply 8 by divisor 12: 8 x 12 = 96. Put the product 96 under the 98.

Subtract the 96 from 98 above it, resulting in the **remainder of 2**: 98 - 96 = 2.

Bring down the next part of the dividend, 7 tens.

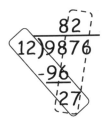

How many 12s fit in 27? Remember 12 x 2 = 24, which is 3 less than 27. So 12 goes into 27 about 2 times: 27/12 = 2 plus a **remainder**. Put the 2 above the 7 in the 10s place.

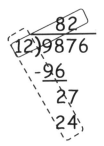

To find the **remainder**, multiply 2 by divisor 12: 2 x 12 = 24. Put the 24 under the 27.

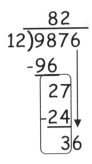

Subtract the 24 from 27 above it, resulting in the **remainder of 3**: 27 - 24 = 3.

Bring down the last part of the dividend, 6 ones.

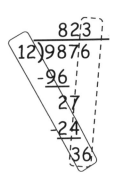

How many 12s fit in 36? Divide 12 into 36. Remember 12 x 3 = 36 exactly. So 36/12 = 3 with **no remainder**. Put quotient 3 above the 6 in the 1s place. All digits in the dividend have been divided into. We are done. (If we multiply 3 x 12 = 36 and place 36 under the 36 and subtract, we get 0.)

Therefore, to earn $9,876 in a year, you must earn $823 each month, or:

9,876 dollars / 12 months = 823 dollars/month

Now you wonder how much you would have to earn each day.

Continue Example: To earn $9,876 in a year, what must you earn on average each day?

Since there are 365 days in a year, we must divide:

$$9,876 \text{ dollars} / 365 \text{ days} = ? \text{ dollars/day}$$

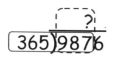

Divide divisor 365 into leftmost part of the dividend it goes into. We see 365 will not "fit" into 9 or 98. So, how many times will 365 go into 987?

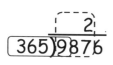

To find how many 365s fit into 987, use the **guess and multiply** strategy. You try 3, but find 365 x 3 = 1,095. Too large. So you try 2. 365 x 2 = 730. Yes, that fits. So 987/365 = 2 plus a **remainder**. Put 2 over the 7 in 987.

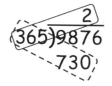

To find the **remainder**, multiply 2 by divisor 365: 2 x 365 = 730.
Put the 730 under the 987.

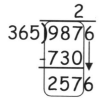

Subtract product 730 from 987 above it, resulting in **remainder 257**: 987 - 730 = 257
Bring down last part of the dividend, 6 ones.

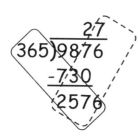

How many 365s fit in 2,576? You know it goes in less than 10 times. Let's try 7: 365 x 7 = 2,555, which is just less than 2,576. Let's use 7. So 365 divides into 2,576 about 7 times: 2,576/365 = 7 plus a **remainder**. Put 7 above the 6.

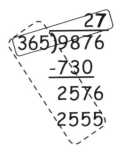

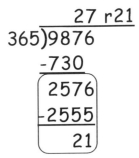

To find the **remainder**, multiply 7 by divisor 365: 7 x 365 = 2,555 (Remember to carry when multiplying.) Put the 2,555 under 2,576.

Subtract 2,555 from 2,576 above it, resulting in **remainder "r" 21**: 2,576 - 2,555 = 21. There are no more numbers in dividend to bring down so write **remainder** r21 next to quotient.

So to earn $9,876 per year, each day you must earn just over 27 dollars. We can write the quotient to the division as:

9,876 dollars / 365 days = 27 21/365 dollars/day

Note: If we divide 21/365 we get about 0.06, or 6 cents, so:

$9,876/365 days is approximately $27.06 dollars/day

(We'll learn about numbers right of the decimal point in Chapter 6.)

Example: What is 60,606,060 ÷ 3,030?

This looks difficult, but with your new tools it's really easy.

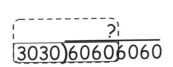

Divide divisor 3,030 into the leftmost part of the dividend it goes into. 3,030 will not go into 6 or 60 or 606, but will go into 6,060. How many times can 3,030 go into 6,060?

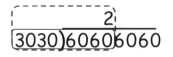

To find how many 3,030s fit into 6,060, look at these numbers carefully. If you double 3,030 you get 6,060! 3,030 x 2 = 6,060. So 6,060/3,030 = 2 exactly, **no remainder**. Put 2 over 0 in 10,000s place.

Multiply 2 by 3,030: 2 x 3,030 = 6,060. Put the 6,060 under 6,060.

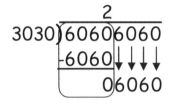

Subtracting 6,060 from 6,060 above it, shows **zero remainder**: 6,060 - 6,060 = 0. Bring down next dividend numbers until you have something 3,030 fits into. 3,030 won't go into 6 or 60 or 606, so bring down all of 6,060.

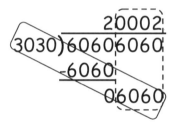

How many 3,030s fit in 6,060? We showed exactly 2 since 2 x 3,030 = 6,060. This says 6,060/3,030 = 2. Put 2 above the 0 ones, filling in 0s in between.

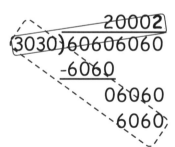

Multiply 2 by 3,030: 2 x 3,030 = 6,060. Put the 6,060 under 6,060.

$$
\begin{array}{r}
20002 \\
3030\overline{)60606060} \\
-6060 \\
\hline
06060 \\
-6060 \\
\hline
0
\end{array}
$$

Subtracting 6,060 from 6,060 above it, shows **zero remainder**: 6,060 - 6,060 = 0 Nothing left to bring down. We're done! **Therefore,**

   60,606,060/3,030 = 20,002

See - this one was fun and easy!

3.3. Divisibility, Factors, and Multiples

Divisibility

What does divisibility of a number mean? It means a number can be divided "evenly" or exactly by another number.

Let's look at the divisibility of 8 by 4:

8 is divisible by 4, since 8 can be
divided by 4 exactly 2 times.
Also, 8 can be divided by 2 four times.

8 is divisible by 4 since 4 goes into 8 exactly 2 times (no remainder).

8 is divisible by 2 since 2 goes into 8 exactly 4 times (no remainder).

Example: What is the divisibility of 12 by 3?

12 is divisible by 3 since 12 can be
divided by 3 exactly 4 times. Also, 12
can be divided by 4 three times.

12 is divisible by 3 since 3 goes into 12 exactly 4 times (no remainder).

12 is divisible by 4 since 4 goes into 12 exactly 3 times (no remainder).

Example: Is 12 divisible by 6?

12 is divisible by 6 since 12 can be
divided by 6 exactly 2 times.
Also, 12 can be divided by 2 six times.

12 is divisible by 6 since 6 goes into 12 exactly 2 times (no remainder).

12 is divisible by 2 since 2 goes into 12 exactly 6 times (no remainder).

What if a number cannot be divided exactly by a second number and leaves a remainder? Then the first number is **not divisible** by the second number. A **remainder** is the amount left over if you divide a number that does not divide exactly by a second number. Remainders result when numbers are not perfectly divisible.

Example: Do you think 8 is divisible by 3?

   3 divides into 8 stars 2 whole times
                                with a remainder of 2.

3 goes into 8 stars 2 whole times with 2 stars **remaining**.
8 is not divisible by 3 since 3 **does not divide exactly** into 8.

Example: Is 12 divisible by 5?

   5 divides into 12 stars 2 whole times
                                plus a remainder of 2.

12 stars are not divisible by 5, since they **cannot be divided exactly** by 5.

Extra Credit Example: Is 50 divisible by 100?

This asks if 100 divides exactly into 50. **No**, 100 is bigger than 50. **100 can't go into 50 an exact number of times**. If you try dividing 50 by 100 you get: **50 ÷ 100 = 50/100 = 0 plus a remainder of 50**. You can write the remainder as 50/100. So, 50/100 = 0 + 50/100 or just 50/100. Since 50/100 = 1/2, we see that 1/2 of 100 goes into 50.

Factor

What is a factor? It is a smaller number that can be divided exactly into a larger number without a remainder.

A smaller number is a **factor** of a larger number if the smaller number divides into the larger number **without** leaving a **remainder**.

Numbers that are multiplied together to produce a product are factors of that product.

Let's look at some factors of 8:

2 and 4 are factors of 8
since 2 x 4 = 8 and 4 x 2 = 8
and 2 and 4 divide exactly into 8.

Example: Show the different factors of 12?

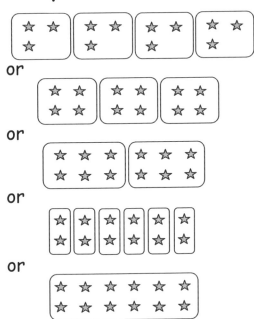

or

3 and 4 are factors of 12
since 3 x 4 = 12 and 4 x 3 = 12
and 3 and 4 divide exactly into 12.

or

2 and 6 are factors of 12
since 2 x 6 = 12 and 6 x 2 = 12
and 2 and 6 divide exactly into 12.

or

1 and 12 are factors of 12
since 1 x 12 = 12 and 12 x 1 = 12
and 1 and 12 divide exactly into 12.

So factors of 12 are: 1, 2, 3, 4, 6, and 12. Note: 5, 7, 8, 9, 10, 11 are **not** factors of 12 since factors divide exactly with no remainder (for example, 12/5 = 2 + remainder 2).

Multiples

What is a multiple of a number? The product of that number times an integer. In other words, a multiple of a number is the number that results after you multiply a number by an integer.

Let's look at multiples of 2:

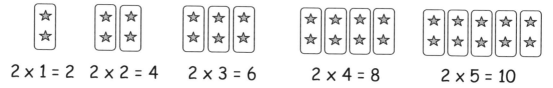

$2 \times 1 = 2$ $2 \times 2 = 4$ $2 \times 3 = 6$ $2 \times 4 = 8$ $2 \times 5 = 10$

We could continue to show multiples of 2 to infinity!

The product of a number and integer is a multiple of the number.

Example: Show some multiples of 4:

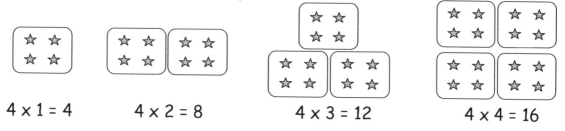

$4 \times 1 = 4$ $4 \times 2 = 8$ $4 \times 3 = 12$ $4 \times 4 = 16$

We could continue to show multiples of 4 to infinity!

You can create infinite multiples of a number by multiplying it by other numbers.

Example: Show multiples of 0.

Uh oh - multiples of zero are just 0! Why? Because any number multiplied by zero, including zero, is just zero.

$0 \times 0 = 0$, $0 \times 1 = 0$, $0 \times 2 = 0$, $0 \times 3 = 0$, $0 \times 4 = 0$, and so on.

3.4. Multiplying and Dividing with Negative Numbers

When you multiply or divide numbers with negative signs, you can first perform the calculation. Then figure out the sign.

Look at the simple example of: 2 x 2   we know   2 x 2 = 4

What if one of the 2's is negative?

2 x -2 = -4 or -2 x 2 = -4

One negative sign makes the answer negative.

Let's look at 2 x -2 = -4 or -2 x 2 = -4 on the number line.

Since the order of multiplication doesn't matter:

2 x -2 = -4 is the same as -2 x 2 = -4

On the number line we can see that
-2 times +2 is -4 and +2 times -2 is -4

positive 2 taken negative 2 times

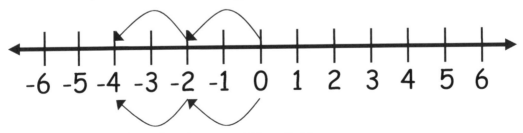

negative 2 taken positive 2 times

What if we multiply two negative numbers?   − x − = ?

What if both 2's are negative? -2 x -2 = 4

Multiplying two negative signs makes the answer positive!

The second negative sign cancels out the first negative sign.

Each time we multiply by a negative, we take the opposite.

When you multiply or divide numbers with negative signs, you first do the calculation. Then figure out the sign. For -2 x -2, we first say 2 x 2 = 4. To get the sign, we first have a -2 which says we go left on the number line. But then the **second negative sign goes the opposite** direction, or right, which is positive.

-2 x -2 means negative 2 negative 2 times

The second - sign canceled the first - sign resulting in a + direction.

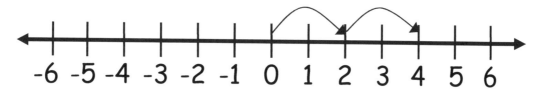

Why does multiplying by a negative take the opposite? Let's look using number lines.

Multiplying by a "-" sign

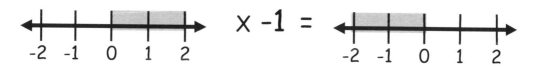

Multiplying by a **negative sign** reverses + **to** − or − **to** +
Let's see how this works...

Multiply Two Positives gives Positive: +2 x +1

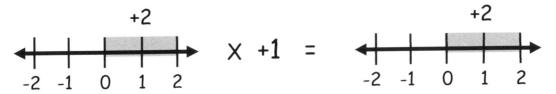

We see multiplying **+** by **+** leaves **+2** as **+2**

Multiply Positive and Negative gives Negative:

+2 x –1= –2 or –2 x +1= –2

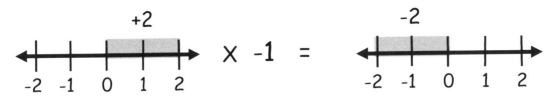

We see multiplying **+** by **–** reverses **+2** to **–2**

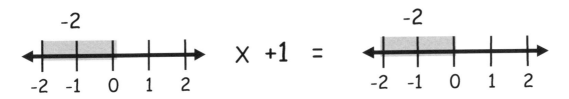

We see multiplying **–** by **+** leaves **–2** as **–2**

Multiply Two Negatives gives Positive: –2 x –1 = +2

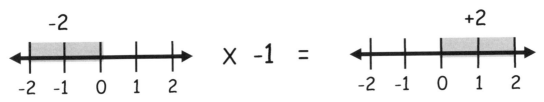

We see multiplying **–** by **–** reverses **–2** to **+2**

With division you divide the numbers and determine the sign.

Divide Positive by Positive gives Positive: +2 ÷ +1 = +2

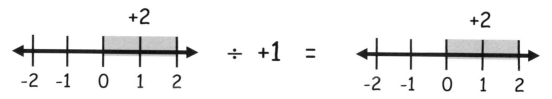

+1 goes into +2 exactly +2 times. The +2 remains +2.

Divide Negative by Positive gives Negative: -2 ÷ +1 = -2

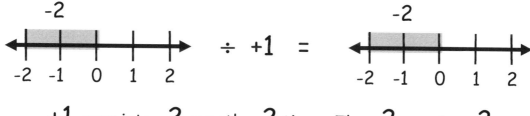

+1 goes into -2 exactly -2 times. The -2 remains -2.

Divide Positive by Negative gives Negative: +2 ÷ -1 = -2

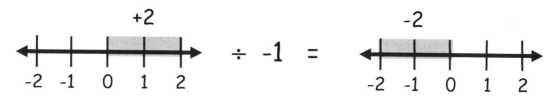

-1 goes into +2 exactly -2 times. The +2 reverses to -2.

Divide Negative by Negative gives Positive: -2 ÷ -1 = +2

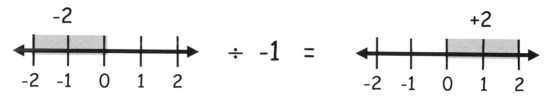

-1 goes into -2 exactly +2 times. The -2 reverses to +2.

What if you multiply (or divide) 3 numbers? Let's see:

Multiply -2 x -1 x -1 = -2

See that multiplying 3 – signs results in a –, since
two – signs give a +, but a third – flips + to –.

To see this first show **-2 x -1 = +2** which reverses -2 to +2

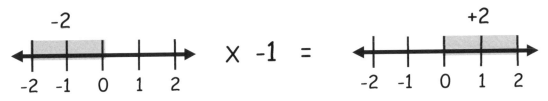

Then show **+2 x -1 = -2** which reverses +2 to -2

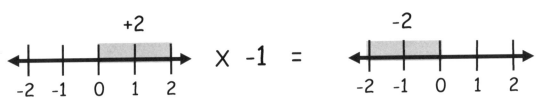

Multiply +2 x -1 x -1 = +2

Multiplying 1 + sign and 2 – signs results in a +, since the two – signs
give a +, and the one + keeps it positive.

To see this first show **+2 x -1 = -2** which reverses +2 to -2

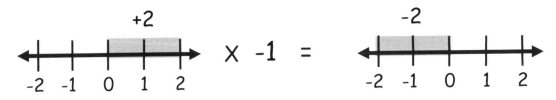

Then show **-2 x -1 = +2** which reverses -2 to +2

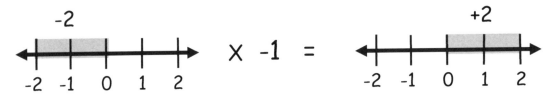

Multiply +2 x +1 x −1 = −2

Multiplying 2 **+** signs and 1 **−** sign results in **−**, since

two **+** signs result in a **+**, but the one **−** flips to negative.

To see this first show **+2 x +1** = **+2** which leaves +2 as +2

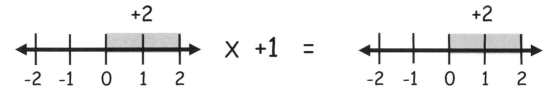

Then show **+2 x −1** = **−2** which reverses +2 to −2

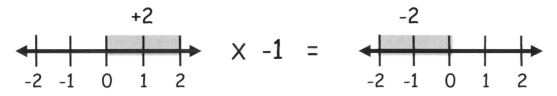

Multiply −2 x +1 x −1 = +2:

This again is multiplying 1 **+** sign and 2 **−** signs which results in a **+**,

since two **−** signs give a **+**, and the one **+** keeps it positive.

To see this first show **−2 x +1** = **−2** which leaves −2 as −2

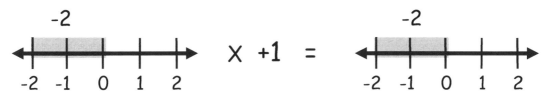

Then show **−2 x −1** = **+2** which reverses −2 to +2

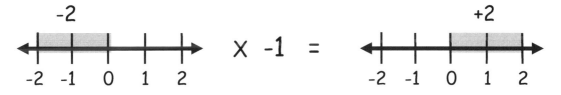

Note: multiplying all positives gives positive: +2 x +1 x +1 = +2

Example: What is -4 x -2 x 3 x -5 x 4?

When you multiply or divide numbers with negative signs, first perform the calculation. Then figure out the sign. Here's the calculation without the signs:

4 x 2 x 3 x 5 x 4

8 x 3 x 5 x 4

24 x 5 x 4

120 x 4 = 480

To find the sign, we started with a: -4.
Then multiplied -4 by -2, which flips the - sign to a + sign: +8.
Then multiplied +8 by +3, which keeps sign +: +24.
Then multiplied +24 by -5, which flips the + sign to a - sign: -120.
Then multiplied -120 by +4, which keeps the - sign: -480.

But there is a faster way to find the sign! Just count negatives.
An even number of - signs gives a +
An odd number of - signs gives a -

Since there are 3 - signs in this example, we get a negative answer. Therefore, our final answer is: -480.

You may be wondering why you care about multiplying and dividing negative numbers. There are many applications which arise in fields of science, engineering, business, etc. Examples include changes in rates, temperatures, altitudes, money debts, and directions.

3.5. Practice Problems

3.1

(a) You must add 7 + 7 + 7 + 7 + 7 + 7 + 7 + 7 + 7 = ? Write this as a multiplication problem using different kinds of notation and then find the product in the multiplication table.

(b) Show 2 x 6 and 6 x 2 on a number line.

(c) You have 20 ft of fence to enclose a garden area. How much bigger will your garden be in square feet if you make it square compared to making it only 3 feet wide?

(d) You purchase 742 lottery tickets for $10 each. How much do you spend? If none of your tickets are winners, how much do you win?

(e) A rectangular farm is 2,000 meters by 4,000 meters. How many square meters does it contain?

(f) Are these all equivalent to each other? Which is not?
$(1 + 2)(3 + 4)$, $(3 + 4)(2 + 1)$, $(1 + 2 + 4)(3)$, $(4 + 3)(1 + 2)$, $(4)(1 + 2 + 3)$

(g) If each shoe has 14 cleats, how many cleats take the field with a team of 11 football players?

(h) In **Section 3.1**, under **Multiplying 3 Numbers**, we showed a flower garden that is 5 feet by 3 feet, with each square foot having 4 flowers. If each flower has 8 petals, how many petals are in the garden?

(i) If you assume Methuselah lived exactly 969 years, how many days did he live (ignore Leap Years)?

(j) If x, y, and z are odd numbers, is the following even or odd?
$(x + y + z)(x * y * z)(x + y + 1)$

3.2

(a) You and 5 friends play a card game. If you deal a 52-card deck so each person gets the same number of cards, how many cards would each player get? How many cards would be left over? What if you added 2 jokers to the deck?

(b) You are having a big party, and you bake ten 32 oz pizzas. You suspect each guest will consume 12 oz of pizza. How many guests can you serve?

(c) Your friend says she is thinking of a number "x" and asks you to solve the following division problems:
$x/1 = ?$, $x/x = ?$, $x/0 = ?$, and x divided by half x.
Can you solve her four problems?

(d) You have 5 children and want to divide $21 equally among them. What should you do?

(e) Your hybrid vehicle traveled 50 miles. You check the gasoline in the tank before and after the trip. You discover that the car used no gasoline and ran entirely on its battery. What miles per gallon (mpg) of gas did you achieve?

(f) How can you use the multiplication tables to find $67 \div 7$?

(g) Use long division to find $4{,}580{,}247 \div 7$.

(h) Use long division to find $15{,}542/38$.

(i) Use long division to find $699{,}678 \div 1{,}234$

3.3

(a) Which of these numbers is divisible by 2?
2,467,   42,501,   107,370,   71,149

(b) Which of these numbers is divisible by 3?
4,153, 29,010, 71,333, 111,111.

(c) Prime Numbers are numbers that are only divisible by themselves and by 1. These include 2, 3, 5, 7, 11, 13, 17, 19, 23, 29, 31, 37, 41, 43, 47, etc. What are the Prime Factors of the number 60? What are ALL the factors of the number 60?

3.4

(a) Find: 48/-8 = ?, -48/8 = ?, and -48/-8 = ?
Check your answers with multiplication.

(b) If "+" means a positive number, "-" means a negative number, and * means multiply, is the following positive or negative?
(+ * - * +)/(- * - * -)

Answers to Chapter 3 Practice Problems

3.1

(a) 7 times 9 = 7 x 9 = 7 * 9 = (7)(9) = 63.

(b)

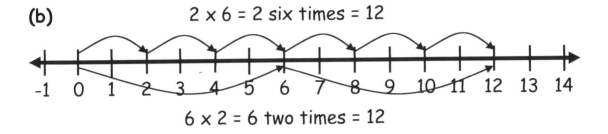

2 x 6 = 2 six times = 12

6 x 2 = 6 two times = 12

(c) A square garden would be 5 ft by 5 ft, having 4 sides of 5 ft each: 5 + 5 + 5 + 5 = 20 ft. Area is: 5 x 5 = 25 square feet. A garden 3 ft wide would be 7 ft long, with the sum of the sides as: 3 + 3 + 7 + 7 = 20 ft. Area is: 3 x 7 = 21 square feet. You could count the squares:

$$5 \begin{cases} 1 & 2 & 3 & 4 & 5 \\ 6 & 7 & 8 & 9 & 10 \\ 11 & 12 & 13 & 14 & 15 \\ 16 & 17 & 18 & 19 & 20 \\ 21 & 22 & 23 & 24 & 25 \end{cases}$$

5

$$3 \begin{cases} 1 & 7 & 8 & 9 & 10 & 11 & 12 \\ 8 & 9 & 10 & 11 & 12 & 13 & 14 \\ 15 & 16 & 17 & 18 & 19 & 20 & 21 \end{cases}$$

7

So the square garden would be larger by 25 – 21 = **4 square feet**. Note: If you used the fencing to make a garden 10 ft long and 0 ft wide, it would have 10 x 0 = 0 square feet. The largest garden you can make with 20 ft of fence is circular and would contain almost 32 square feet!

(d) To multiply an integer by 10 you can move each digit to the left one place by putting a zero at the end (in the ones place). So $742 \times 10 = \$7,420$. If all tickets are losers, your winnings are $742 \times \$0 = \$0$. You can't get your \$7,420 back either!

(e) You could use column form to calculate this, but a quick path to the answer is to think of it as a plot $2 \times 1,000$ m wide and $4 \times 1,000$ m long. Using the associative property, this can be written as $(2 \times 4) \times (1,000 \times 1,000) = (8)(1,000)(1,000)$. Remember when we multiply a **whole number** such as 2 by $1,000$, we can put the 3 zeros to the right of the 2. This actually moves the 2 digit 3 places to the left, which results in $2,000$. Multiplying the 8 by $(1,000)(1,000)$ moves the 8 digit 6 places to the left. You do this by inserting 6 zeros to the right of the 8. So $2,000 \times 4,000 = 8,000,000$, or 8 million square meters, or 8 square kilometers.

(f) The first four expressions all equal 21, but
$(4)(1 + 2 + 3) = (4)(6) = 24$

(g) Each player has 2 shoes, so each player wears 14×2 cleats.

$$\begin{array}{r} 14 \\ \times 2 \\ \hline 8 \end{array} \qquad 4 \times 2 = 8 \qquad \begin{array}{r} 14 \\ \times 2 \\ \hline 28 \end{array} \qquad \begin{array}{l} 1 \times 2 = 2 \\ \text{so } 14 \times 2 = 28 \text{ cleats} \end{array}$$

Since each player has 28 cleats, 11 players have 28×11 cleats.

$$\begin{array}{r} 28 \\ \times 11 \\ \hline 8 \end{array} \qquad \begin{array}{r} 28 \\ \times 11 \\ \hline 28 \end{array} \qquad \begin{array}{r} 28 \\ \times 11 \\ \hline 28 \\ 8 \end{array} \qquad \begin{array}{r} 28 \\ \times 11 \\ \hline 28 \\ 28 \\ \hline 308 \end{array} \qquad \begin{array}{l} \textbf{308 cleats} \\ \textbf{take the} \\ \textbf{field!} \end{array}$$

(h) What we are asking is 5 x 3 x 4 x 8 = ? We already learned that 5 x 3 x 4 = 60. So we need to find: 60 x 8 = ? We can just say 6 x 8 = 48 and insert the zero after the 8 to get 480. This is just moving the 48 one place to the left while putting 0 in the 1s place to give 480 petals.

(i) Methuselah lived 969 years x 365 days/year. Let's multiply:

3 4		4 5		2 2	
969	first	969	second	969	third
x365	partial	x365	partial	x365	partial
4845	product	4845	product	4845	product
		5814		5814	
				2907	
				353685	

Methuselah lived about **353,685 days**. Just for fun, since there are 24 hours/day x 60 minutes/hour x 60 seconds/minute = 86,400 seconds/day, we see he lived about 353,685 days x 86,400 sec/day, which equals about 30,558,384,000 seconds. It's OK to use a calculator for big calculations like this!

(j) 3 odd numbers added together sum to an odd number, so (x + y + z) is odd. When 3 odd numbers are multiplied together, their product is odd, so (x * y * z) is odd. Two odd numbers (x + y) sum to an even number, but then adding 1 makes an odd number, so (x + y + 1) is odd. Finally multiplying the 3 odd numbers in the 3 sets of parentheses also gives an odd number, so the final result is odd. (Check this by substituting numbers.)

3.2

(a) 52/6 = ? From the multiplication tables you remember that 6 × 8 = 48. So 52/6 = 8 r 4. Therefore each player could get 8 cards with 4 cards left over. Adding 2 jokers would increase the remainder to 6, so each player could get one more card. Or we could just write: 54/6 = 9 with no remainder.

(b) 32 oz per pizza times 10 pizzas = 32 oz × 10 = 320 total oz of pizza. Now determine how many times does 12 oz fit in 320 oz:

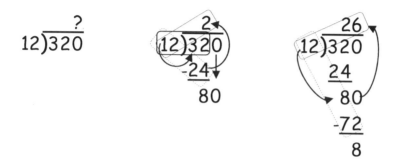

You can serve **26 guests** with 8 oz left for the dog. Note that the 8 oz remainder is: 8 oz / 32 oz pizza = 1/4 of a whole pizza.

(c) Yes, you can solve all of them!
Any number divided by 1 is the number itself, so x/1 = x.
A number divided by itself is 1, so x/x = 1.
A number divided by 0 is undefined, so x/0 is undefined.
Finally, half of any number fits into the number 2 times,
so x ÷ half x = 2. For example 100 ÷ 50 = 2, 6 ÷ 3 = 2, one
divided by one half = (1/½) = 2.

(d) 21 ÷ 5 = ? Since 5 x 4 = 20, then 20/5 = 4. 21 - 20 = 1.
If you give each child $4, you would have $1 left over:
21 ÷ 5 = 4 r 1. So how could you divide the remaining $1 equally?
$1 = 100 cents. 100 ÷ 5 = ?

$$
\begin{array}{r}
20 \\
5\overline{)100} \\
-10\downarrow \\
\hline
00
\end{array}
$$

If you changed the extra dollar into dimes, each child would get
an additional 2 dimes or 20 cents, so $21 ÷ 5 = **$4.20**.

(e) Miles per gallon = miles traveled divided by gallons used.
50 miles / 0 gallons = undefined. When no gasoline is used, the
concept of miles per gallon is not applicable. Note: If you had
used one millionth of a gallon, you would have achieved 50 million
mpg, a large number, but at least it is defined!

(f) You look for the largest number in the 7s column which does
not exceed 67. You find 63 in the row across from 9.
67 – 63 = 4. Therefore, 67/7 = 9 r 4, or 9 4/7.

(g) Using long division to find 4,580,247 ÷ 7 = ?

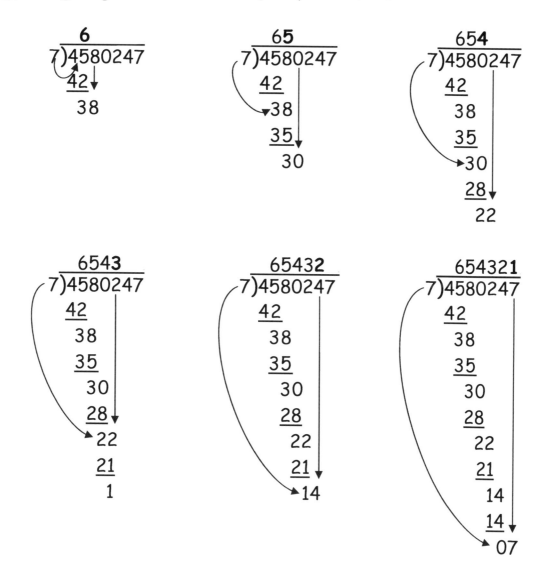

Therefore, 4,580,247/7 = **654,321**.

(h) Use long division to find 15,542/38 = ?

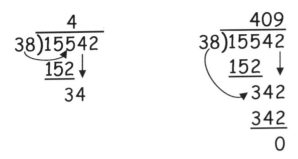

Therefore, 15,542/38 = **409**.

(i) Use long division to find 699,678/1,234 = ?

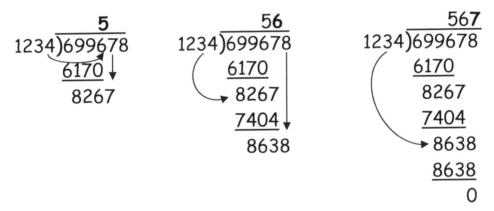

Therefore, 699,678/1,234 = **567**.

3.3

(a) You can divide each number by 2 to see which give a remainder of 0, which would mean it is divisible by 2. An easier way is to see which numbers have an even number in the 1s place. The only number with an even 1s place is the one ending in 0. So 107,370 must be divisible by 2. 107,370/2 = 53,685 with no remainder. This shortcut works because any number of 10s, 100s, 1,000s, etc. is divisible by 2 (since 10/2 = 5 with no remainder).

(b) You can divide all the numbers by 3 to see which have a remainder of 0. **A shortcut is to add all the digits in each number. If the sum of the digits is divisible by 3, then the original number is also divisible by 3.** Let's look:

$4 + 1 + 5 + 3 = 13$ and $13/3 = 4$ r1, so 4,153 is **not** divisible by 3.

$2 + 9 + 0 + 1 + 0 = 12$ and $12/3 = 4$, so 29,010 **is** divisible by 3. $(29,010/3 = 9,670)$

$7 + 1 + 3 + 3 + 3 = 17$ and $17/3 = 5$ r2, so 71,333 is **not** divisible by 3.

$1 + 1 + 1 + 1 + 1 + 1 = 6$ and $6/3 = 2$, so 111,111 **is** divisible by 3. $(111,111/3 = 37,037)$

(c) One way to find the prime factors of a number is to divide it by the <u>prime numbers</u> that divide into it exactly, beginning with 2, then moving to 3, then to 5, then to 7, etc. until the quotient becomes 1.

$60 \div 2 = 30$ So one factor is 2

$30 \div 2 = 15$ So another factor is a second 2

$15 \div 3 = 5$ So another factor is 3

$5 \div 5 = 1$ So the final factor is 5

We see the Prime Factors of 60 are $2 \times 2 \times 3 \times 5 = 60$.

The **factors** of 60 are ALL the numbers that divide exactly into 60. These include both 1 and 60, as well as the Prime Factors 2, 3, and 5. **All** of the factors of 60 include the products obtained by multiplying the Prime Factors which are less than 60:

$4 = (2 \times 2)$, $6 = (2 \times 3)$, $10 = (2 \times 5)$, $12 = (2 \times 2 \times 3)$,

$15 = (3 \times 5)$, $20 = (2 \times 2 \times 5)$ and $30 = (2 \times 3 \times 5)$.

So, all factors of 60 are: 1, 2, 3, 4, 5, 6, 10, 12, 15, 20, 30, and 60.

Note how they pair off to equal 60:

$1 \times 60 = 2 \times 30 = 3 \times 20 = 4 \times 15 = 5 \times 12 = 6 \times 10 = 60$

3.4

(a) To find 48/-8, first find 48/8 and then take the negative. Using the multiplication table or remembering that 6 x 8 = 48, you find 48/8 = 6. The negative of 6 is -6, so **48/-8 = -6.** Check: -6 x -8 = 48? First we know 6 x 8 = 48. Then, since both -8 and -6 are negative, take the negative **two** times. The first negative sign makes 48 become -48. The second negative sign makes -48 become 48, giving the original multiplicand 48.

To find -48/8, we know 48/8 = 6. Then take the negative: -6, so **-48/8 = -6.** Check: -6 x 8 = -48? We know 6 x 8 = 48. Then take the one negative for -6, which gives: -48.

To find -48/-8, we know 48/8 = 6. Take the negative twice. The first negative sign gives -6. The second negative gives +6. So, **-48/-8 = 6.** Check: 6 x -8 = -48? We know 6 x 8 = 48. Taking the negative gives -48.

(b) You can first determine if the dividend and divisor are positive or negative and then determine if their quotient is positive or negative. Remember, a product is negative only if there are an odd number of negative factors.

(+ * - * +) is negative because there is 1 negative factor.

(- * - * -) is negative because there are 3 negative factors.

Now dividing a negative number by a negative number (-/-) gives a positive quotient, so the answer will be positive.

A quicker way to determine if the answer is positive or negative: Count the number of **negative** numbers in both the dividend and divisor (also called the numerator and denominator). You find a total of 4 negative numbers, giving an even number of negatives. That tells you the answer will be positive.

Chapter 4

Fractions and How to Add and Subtract Them

*The Lord is my **portion**, saith my soul; therefore will I hope in Him. Lamentations 3:24*

A fraction is a part of a whole:

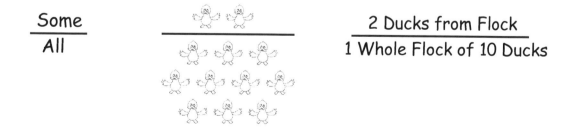

$$\frac{\text{Part}}{\text{Whole}} \qquad\qquad \frac{\text{1 Duck from Flock}}{\text{1 Whole Flock of 10 Ducks}}$$

A fraction is some of the all:

$$\frac{\text{Some}}{\text{All}} \qquad\qquad \frac{\text{2 Ducks from Flock}}{\text{1 Whole Flock of 10 Ducks}}$$

4.1. What are Fractions?

A fraction is a part of the whole. Think of a pie:

$$\frac{\text{Part}}{\text{Whole}}$$
　　　　　　　　　　　　$$\frac{\text{1 Piece of the Pie}}{\text{Whole Pie}}$$

The fraction expresses how large the part is compared to the whole.

A pie can be cut into fractions:

If a pie is cut into 8 equal pieces,
3 pieces are part of the whole 8 pieces.

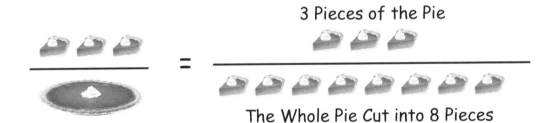

3 Pieces of the Pie

The Whole Pie Cut into 8 Pieces

Together the 3 pieces are 3 of the whole 8 pieces

A group of balls can be described as a fraction:

If you have 5 balls total,
4 balls are a part of your 5 total balls.

$$= \frac{\text{4 of the 5 Balls}}{\text{The Total 5 Balls}} = \frac{\text{Part of a Group}}{\text{The Whole Group}}$$

Fractions are used in our daily lives as we divide a pizza or a group of objects into halves (of the whole), thirds (of the whole), fourths (of the whole), etc. When we divide into halves, thirds, fourths, etc, the pieces or groups are **always of equal size**. Let's see:

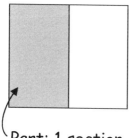

Halves
2 equal-size
pieces or groups

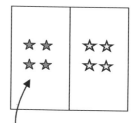

Part: 1 section
Whole: 2 sections

Part: 4 stars
Whole: 8 stars

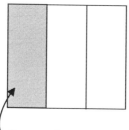

Thirds
3 equal-size
pieces or groups

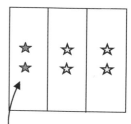

Part: 1 section
Whole: 3 sections

Part: 2 stars
Whole: 6 stars

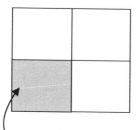

Fourths, or quarters
4 equal-size
pieces or groups

Part: 1 section
Whole: 4 sections

Part: 4 stars
Whole: 16 stars

How are fractions written?

Fractions are written: 3/4 or $\dfrac{3}{4}$ or $\dfrac{\text{number of parts}}{\text{number of parts making the whole}}$

Using letters x and y to represent numbers, we write: x/y or $\dfrac{x}{y}$

What are the numbers called: $\dfrac{\text{top number is the } \textbf{numerator}}{\text{bottom number is the } \textbf{denominator}}$

Fractions Express Division!

$1/2 = \dfrac{1}{2} = 1 \div 2$ The division $1 \div 2 = 1/2$
shows half of the whole 2 fits into the 1.

Let's review division and fractions as they are defined:

Division shows how many times the bottom number fits into the top number. **Writing division as a fraction shows how many "wholes" (bottom) fit into the "part" (top).**

In a **fraction**, the **top** number is how many **parts** and the **bottom** number is how many parts make up the **whole**.

$$\text{Fraction} = \dfrac{\text{Part}}{\text{Whole}} = \text{Division}$$

Can We See Fractions on a Number Line? Yes!

Example: Show the fraction 1/2 on a number line:

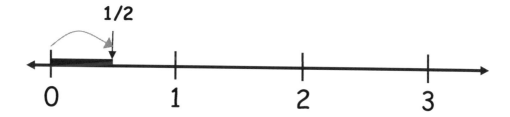

This is 1/2 of a whole 1.

Example: Show the fraction 3/4 on a number line:

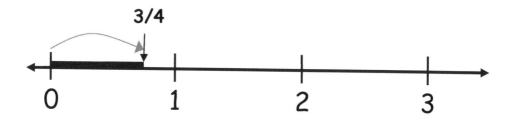

This is 3/4 of a whole 1.

Example: Show the fraction -1/2 on a number line:

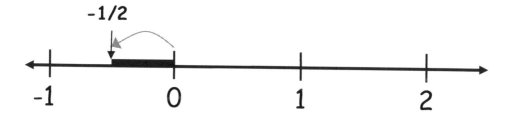

This is 1/2 of a whole -1.

Equivalent Fractions

Some fractions have the same size or value.
They are called **equivalent fractions.**

The following are equivalent fractions. They are all equal!

1/2   =   2/4   =   4/8   =   8/16

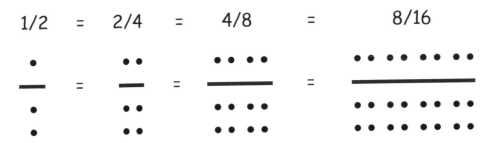

See that the **whole** (bottom) has twice the dots as the **part** (top).

Example: Show equivalent fractions where the denominator (bottom) value is twice the numerator (top) value.

1/2   =   4/8   =   16/32

Example: What are the first 7 equivalent fractions of 1/2 where both the numerator and denominator are doubling?

1/2 = 2/4 = 4/8 = 8/16 = 16/32 = 32/64 = 64/128

We can also see equivalent fractions as:

1/2   =   2/4   =   8/16

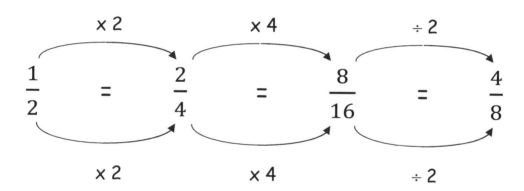

How Do You Find Equivalent Fractions?

Equivalent fractions are found by multiplying or dividing the numerator (top) and denominator (bottom) by the same values. This keeps their relative size the same. In the following example, each denominator is twice the size of its numerator.

To keep a fraction equivalent, so its value stays the same, when you multiply or divide the top by a number, you must also multiply or divide the bottom by the same number.

Example: Show that each pair of fractions is equivalent:
3/4 = 6/8, 30/40 = 3/4, 2/3 = 6/9, 10/15 = 2/3, 3/5 = 45/75

To show $\dfrac{3}{4}$ is equivalent to $\dfrac{6}{8}$ Multiply top and bottom by 2 $\dfrac{3}{4} \dfrac{\times 2}{\times 2} = \dfrac{6}{8}$

To show $\dfrac{30}{40}$ is equivalent to $\dfrac{3}{4}$ Divide top and bottom by 10 $\dfrac{30}{40} \dfrac{\div 10}{\div 10} = \dfrac{3}{4}$

To show $\dfrac{2}{3}$ is equivalent to $\dfrac{6}{9}$ Multiply top and bottom by 3 $\dfrac{2}{3} \dfrac{\times 3}{\times 3} = \dfrac{6}{9}$

To show $\dfrac{10}{15}$ is equivalent to $\dfrac{2}{3}$ Divide top and bottom by 5 $\dfrac{10}{15} \dfrac{\div 5}{\div 5} = \dfrac{2}{3}$

To show $\dfrac{3}{5}$ is equivalent to $\dfrac{45}{75}$ Multiply top and bottom by 15 $\dfrac{3}{5} \dfrac{\times 15}{\times 15} = \dfrac{45}{75}$

Note: 2/2 = 1, 10/10 = 1, 3/3 = 1, 5/5 = 1, and 15/15 = 1, so multiplying and dividing by these does not change the first fraction's value.

Example: For equivalent fractions 3/5 = 6/?, what is "?"

To make an equivalent fraction, the numerator and denominator of the first fraction must be multiplied by the same number. Look at the top numbers in our example first.

  Numerator 3 is multiplied by **2** to get numerator 6: $3 \times 2 = 6$

Since the numerator and denominator must be multiplied by the same number, we multiply both top and bottom by **2**.

$$\frac{3 \times 2}{5 \times 2} = \frac{6}{10} \qquad \textbf{Therefore, "?" is 10}$$

Example: For equivalent fractions 20/30 = ?/6, what is "?"

The second denominator is smaller, so we must need to divide both the top and bottom by the same number.

Since $30 \div \mathbf{5} = 6$, then: $\dfrac{20 \div 5}{30 \div 5} = \dfrac{?}{6} = \dfrac{4}{6}$ **So, "?" is 4.**

Cross Multiplication is a fun and useful fact about equivalent fractions where cross-multiplied numbers are equal.

$\dfrac{1}{2} \diagtimes \dfrac{2}{4}$ see that

$1 \times 4 = 2 \times 2$

$4 = 4$

$\dfrac{3}{4} \diagtimes \dfrac{6}{8}$ see that

$3 \times 8 = 4 \times 6$

$24 = 24$

4.2. Add or Subtract Fractions with the Same Denominator

Adding or subtracting fractions with the <u>same</u> denominators is like adding cats-with-cats or dogs-with-dogs.

1 cat + 1 cat = 2 cats

$$\text{🐱} + \text{🐱} = \text{🐱🐱}$$

Or

1 dog + 1 dog = 2 dogs

When fractions have the same denominators (bottom numbers), you can just add or subtract their numerators (top numbers) and put the answer over the bottom number.

Let's first look at fractions that have the **same denominators**. The following fractions all have a denominator of 2:

$$\frac{1}{2}, \quad \frac{2}{2}, \quad \frac{3}{2}, \quad \frac{4}{2}$$

Adding or subtracting fractions with the <u>same</u> denominators is adding like-fractions together,
such as **halves-with-halves** or **third-with-thirds**:

$$\frac{1}{2} + \frac{2}{2} = \frac{3}{2} \quad \text{or} \quad \frac{1}{3} + \frac{2}{3} = \frac{3}{3}$$

We can see adding and subtracting fractions by drawing pizzas that have been cut into halves, quarters, and eighths:

Halves:

$$\frac{1}{2} \quad + \quad \frac{1}{2} \quad = \quad \frac{1+1}{2} \quad = \quad \frac{2}{2} = 1$$

Quarters:

$$\frac{1}{4} \quad + \quad \frac{2}{4} \quad = \quad \frac{1+2}{4} \quad = \quad \frac{3}{4}$$

1 fourth + 2 fourths = 1+2 fourths = 3 fourths of whole pizza

The above pizza is cut into 4 equal pieces. Since all the quarter-sized slices are the same size, they can be added together directly.

Now let's look at a couple of examples of eighths where we show
subtraction:

Eighths:

$$\frac{5}{8} \quad - \quad \frac{3}{8} \quad = \quad \frac{5-3}{8} \quad = \quad \frac{2}{8} \quad = \quad \frac{1}{4}$$

$$\frac{5}{8} \quad - \quad \frac{5}{8} \quad = \quad \frac{5-5}{8} \quad = \quad \frac{0}{8} = 0$$

We just saw that if we want to add or subtract fractions
that have the same denominator we can:
1. Combine all the top numbers over the single bottom number.
2. Add or subtract the top numbers.

Look at Adding and Subtracting Fractions on a Number Line

Example: Show 3/4 + 1/4 on a number line:

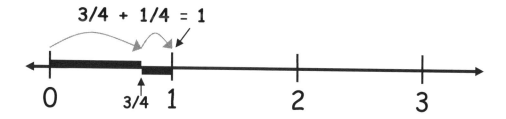

This shows 3/4 + 1/4 = 1.

Example: Show 1/4 - 3/4 on a number line:

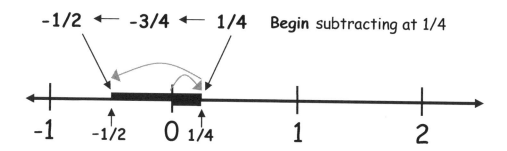

This shows 1/4 - 3/4 = -1/2.

Number lines help us SEE that fractions are points on a number line just like other numbers. But it is usually quicker to add and subtract fractions using the fraction format rather than using a number line. Let's do some addition and subtraction examples with more than two fractions.

Example: Add 2/5 + 3/5 + 4/5.

All denominators are fifths. This has only addition of fifths. Combine and add the top numbers over the bottom number, 5.

$$\frac{2}{5} + \frac{3}{5} + \frac{4}{5} = \frac{2+3+4}{5} = \frac{9}{5}$$

Example: What is 3/7 + 6/7 + 4/7 - 2/7 - 1/7?

All denominators are sevenths. **Add and subtract** top numbers, placing result over single bottom number, 7.

$$\frac{3}{7} + \frac{6}{7} + \frac{4}{7} - \frac{2}{7} - \frac{1}{7} = \frac{3+6+4-2-1}{7} = \frac{10}{7}$$

Example: What is 3/8 + 4/8 - 2/4?

This has **different denominators**, which we will learn about in the next section. You cannot easily subtract the 2/4 because it has a different denominator than the 8ths. But 2/4 can also be written in 8ths by multiplying 2/4 by 2/2. Let's see:

$$\frac{2}{4} \frac{\times 2}{\times 2} = \frac{4}{8}$$

Replace 2/4 = 4/8 so the denominators are all the **same**. Add and subtract the top numbers. Place over the bottom number:

$$\frac{3}{8} + \frac{4}{8} - \frac{4}{8} = \frac{3+4-4}{8} = \frac{3}{8}$$

4.3. Add or Subtract Fractions with Different Denominators

What if you want to add or subtract two fractions but the denominators are different?

To add or subtract fractions with different denominators, you must first make the denominators the same, while adjusting the numerators to keep them equivalent. Then you can add or subtract the new numerators and set the answer over the new denominator. Making the denominator the same puts the fraction into **common units**. We will learn how to do this, but first let's think about **common units**. What if you want to add cats and dogs? We need a common unit.

2 cats +   3 dogs = ?

If we think of cats and dogs as **both being animals**:

2 cats = 2 animals   and   3 dogs = 3 animals

We can now add them: 2 **animals** + 3 **animals** = 5 **animals**

Like animals, to **add** or **subtract** different fractions, such as halves (1/2) and thirds (1/3), we must make **common units**. For fractions, this means the **denominators must be the same.**

Look at 1/4 + 1/3 = ?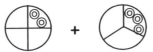

First make the denominators the same "common denominator".
The common denominator will have the 4 from 1/4 and the 3
from 1/3 inside it. It will be a multiple of 4 and of 3.

An easy way to make a new denominator with both 4 and 3 inside
it is to just multiply 4 x 3. In other words:

**A quick way to find a common denominator is to
multiply the two denominators by each other.**

$$\frac{1}{4} + \frac{1}{3}$$
$$\times$$

$$4 \times 3 = 12$$

This means 12 is your new common denominator.

**But you must adjust the numerators to account for the new
denominator.** To do this, **multiply each top number by the
other fraction's original bottom number** (since we already
multiplied its denominator by the other denominator). Let's see:

$$\frac{1}{4} \times \frac{1}{3} \implies \frac{(1 \times 3) + (1 \times 4)}{12} = \frac{3+4}{12}$$

Denominator 4 is multiplied by 3, so its numerator 1 multiplies by 3.
Denominator 3 is multiplied by 4, so its numerator 1 multiplies by 4.

Now we add the new numerators and put the answer over the new
common denominator.

$$\frac{3+4}{12} = \frac{7}{12}$$

Let's summarize how to find a common denominator so
we can add or subtract fractions with different denominators:

To add or subtract two fractions with different bottom numbers:

1. Get a common denominator by multiplying both denominators.

$$\frac{1}{4} + \frac{1}{3}$$

2. Multiply each numerator by the other original denominator.

$$\frac{1}{4} \quad \frac{1}{3}$$

3. Add or subtract new numerators and place sum over the new denominator.

$$\frac{(1 \times 3) + (1 \times 4)}{4 \times 3} = \frac{3 + 4}{12} = \frac{7}{12}$$

4. Reduce final answer if needed.
   (We will discuss reducing fractions in Chapter 5.)

We can show the steps all together:

$$\frac{1}{4} \; + \; \frac{1}{3} = \frac{(1 \times 3) + (1 \times 4)}{4 \times 3} = \frac{3 + 4}{12} = \frac{7}{12}$$

Let's draw 1/4 + 1/3:

1/4 + 1/3

By multiplying the denominators we found that one common denominator is 12. Then we adjusted each numerator by multiplying it by the other original denominator.

We can draw our pizzas to show the new numerators and new common denominator:

3/12   4/12   7/12

We see that: 1/4 + 1/3 = 3/12 + 4/12 = 7/12

Can You Also Show Adding and Subtracting Fractions With Different Denominators on the Number Line? Yes...

Example: Show 1/4 + 1/3 on the number line:

Let's look at 1/4 and 1/3 on two different number lines:

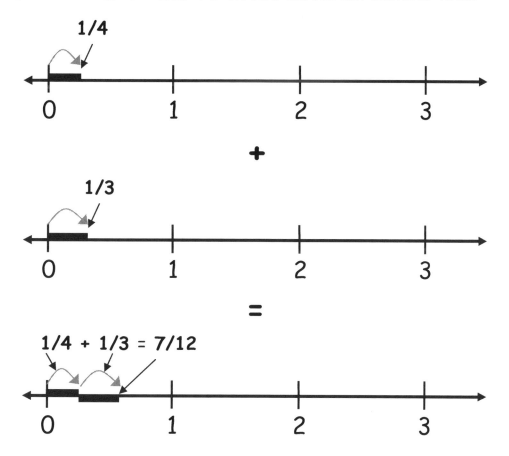

On the previous page we saw **1/4 + 1/3 = 3/12 + 4/12 = 7/12.**

If we divide the number line between 0 and 1 into 12ths, we can **see** that: **1/4 + 1/3 = 3/12 + 4/12 = 7/12**

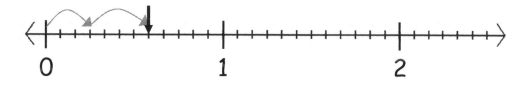

As shown above, we can calculate 1/4 + 1/3 faster than we can draw it, but it is fun to visualize what we are calculating!

Example: What is 3/4 + 4/5?

4ths and 5ths are different sized denominators. We need a common denominator. To find it we:
Multiply the denominators to make a common denominator;
Multiply each numerator by the other original denominator;
Add the new numerators; and
Put the sum over the new common denominator.

$$\frac{3}{4} \times \frac{4}{5} = \frac{(3 \times 5) + (4 \times 4)}{4 \times 5} = \frac{15 + 16}{20} = \frac{31}{20}$$

Therefore, 3/4 + 4/5 = 15/20 + 16/20 = 31/20

Note: $\dfrac{31}{20} = \dfrac{20}{20} + \dfrac{11}{20} = 1\dfrac{11}{20}$ (We'll learn this in Chapter 5.)

Example: What is 3/4 - 2/3?

Multiply the denominators to make a common denominator;
Multiply each numerator by the other original denominator;
Subtract the new numerators; and
Put the **difference** over the new common denominator.

$$\frac{3}{4} \times \frac{2}{3} = \frac{(3 \times 3) - (2 \times 4)}{4 \times 3} = \frac{9 - 8}{12} = \frac{1}{12}$$

Therefore, 3/4 - 2/3 = 9/12 - 8/12 = 1/12

What if you have more than two fractions to add or subtract, and those fractions have different denominators?

One way to add or subtract more than two fractions is to begin by adding or subtracting the first two fractions. Then add or subtract the answer of the first two fractions with the third fraction. If there is a fourth fraction, add or subtract the answer of the first three fractions with the fourth fraction. If there is a fifth fraction, add or subtract the answer of the first four fractions with the fifth fraction. And so on.

Example: What is 1/2 + 2/3 - 1/4?

Add first two fractions, **1/2 + 2/3**:
Multiply denominators to get common denominator. Multiply numerators with other initial denominator. Add numerators. Put over denominator.

$$\frac{1}{2} \times \frac{2}{3} = \frac{(1 \times 3) + (2 \times 2)}{2 \times 3} = \frac{3 + 4}{6} = \frac{7}{6}$$

Subtract third fraction from answer to first two, **7/6 - 1/4**:
Multiply numerators with other initial denominator. Multiply denominators to get common denominator. Subtract numerators. Put over denominator.

$$\frac{7}{6} \times \frac{1}{4} = \frac{(7 \times 4) - (1 \times 6)}{6 \times 4} = \frac{28 - 6}{24} = \frac{22}{24}$$

Our final answer: 1/2 + 2/3 - 1/4 = 7/6 - 1/4 = **22/24**
Note: 22/24 reduces to **11/12** (by dividing top and bottom by 2).

Example: What is 1/2 - 1/4 + 1/6?

Subtract first two fractions, **1/2 - 1/4**:
Multiply numerators with other initial denominator. Multiply denominators to get common denominator. Subtract numerators. Put over denominator.

$$\frac{1}{2} \diagup\kern-1.2em\diagdown \frac{1}{4} \;=\; \frac{(1 \times 4) - (1 \times 2)}{2 \times 4} \;=\; \frac{4 - 2}{8} \;=\; \frac{2}{8}$$

Since 2 divides exactly into the 2 and the 8 in 2/8, we can reduce 2/8 by dividing its top and bottom numbers by 2:

$$\frac{2 \div 2}{8 \div 2} \;=\; \frac{1}{4} \qquad \text{Therefore, } 2/8 = 1/4$$

Now add the third fraction 1/6 to the reduced answer 1/4 of the first two fractions. So we now add: **1/4 + 1/6**:

$$\frac{1}{4} \diagup\kern-1.2em\diagdown \frac{1}{6} \;=\; \frac{(1 \times 6) + (1 \times 4)}{4 \times 6} \;=\; \frac{6 + 4}{24} \;=\; \frac{10}{24}$$

10/24 is our answer, but we can reduce 10/24 by dividing its top and bottom numbers by 2. Let's see the reduced answer:

$$\frac{10 \div 2}{24 \div 2} \;=\; \frac{5}{12} \qquad \text{So, } 10/24 = 5/12$$

Our answer is: 1/2 - 1/4 + 1/6 = 10/24 = 5/12

Another way to add/subtract fractions of different denominators is shown in Master Math: Basic Math and Pre-Algebra Section 2.4.

4.4. Practice Problems

4.1

(a) You have a 32 oz carton of milk. You pour 8 oz into a glass. What fraction of the carton of milk did you pour into the glass?

(b) You cut a candy bar into 2 equal halves. You then cut each half into 2 equal pieces. Finally you cut each piece into 3 equal fragments. You give your sister 5 fragments. What fraction of a whole candy bar does she have?

(c) How many times does 40 fit into 5?

(d) Start with 1/2. Add 1 to the numerator and 1 to the denominator. Write the answer and again add 1 to the numerator and 1 to the denominator. Repeat this process two more times. Are any of the resulting fractions equivalent? Why or why not?

(e) You have 36 eggs and your friend has 48 eggs. You both fill empty egg cartons (12 eggs per carton). What fraction of the eggs do you have? What fraction of the cartons do you have? How can you show that both of your fractions are equivalent?

(f) Sam is 1/8 the age of his grandmother. His grandmother just turned 72. How old is Sam? (Use cross multiplication) What fraction of his grandmother's age will Sam be in 9 more years?

$$\frac{1}{8} \times 72 = \frac{1}{8} \times \frac{72}{1}$$

(g) Your friend brings 15 marbles to "Show and Tell" and says the 15 marbles represent 3/40 of his entire collection. How big is his entire collection?

4.2

(a) A dime is 1/10 of a dollar. You have 5 dimes and your two friends each have 6 dimes. You put them all into a jar. Then you each take out 2 dimes. How many dimes are left in the jar?

(b) A restaurant sells pizza by the slice. Each slice is 1/8 of a whole pizza. The restaurant has 9 warm slices available for sale. A group of 30 pizza lovers shows up and each group member orders 3 slices to go. How many pizzas does the restaurant need to bake to meet the sudden demand? How many slices will be left unsold?

4.3

(a) You have a penny, a nickel, a dime, a quarter, and a half dollar. How much money do you have?

(b) There are 4 pizzas of equal size at a party, each a different flavor you like. One is cut into 6 equal pieces, one into 8 equal pieces, one into 10 equal pieces, and one into 12 equal pieces. You cannot make up your mind, so you eat one slice of each pizza. What fraction of a whole pizza did you eat?

(c) You have 7/8 of a ton of dirt piled on your driveway. Your friend delivers 1/3 ton of dirt, which you add to the pile. You use 2/5 of a ton for your garden. How much dirt is left to spread on your lawn?

Answers to Chapter 4 Practice Problems

4.1

(a) $\dfrac{\text{Glass of Milk}}{\text{Carton of Milk}} = \dfrac{\text{Part of Milk}}{\text{All of Milk}} = \dfrac{8 \text{ oz}}{32 \text{ oz}}$ or $\dfrac{8}{32}$ **of a carton**

Since 8 divides into 8 one time and into 32 eight times:

8/32 = 1/4 or 1/4 of a carton.

Another way to visualize this is to determine how many 8 oz glasses can be filled from a 32 oz carton: $32 \div 8 = 4$
So you can fill 4 glasses. Therefore, the fraction would be:

$\dfrac{1 \text{ glass}}{4 \text{ glasses}} = \dfrac{8 \text{ oz}}{32 \text{ oz}}$ or **1/4 of the carton**

Again, 8/32 and 1/4 are equivalent fractions, so: **8/32 = 1/4**

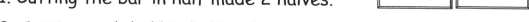

8oz 8oz 8oz 8oz   32oz

(b) First determine how many fragments make up the whole.

1. Cutting the bar in half made 2 halves.

2. Cutting each half in half makes 4 pieces.

3. Cutting each piece into 3 fragments
   makes 12 fragments.

Your sister's fraction of the candy bar is:

$\dfrac{5 \text{ pieces}}{12 \text{ pieces}}$ or **5/12 of the candy bar.**

(c) Less than one time! Written as a division problem, this is
$5 \div 40$ or 5/40. Notice that 40 is a multiple of 5.
Let's reduce 5/40 by dividing top and bottom by 5:

$$\frac{5 \div 5}{40 \div 5} = \frac{1}{8}$$ **So one-eighth of 40 fits into 5.**

(d) $\dfrac{1}{2}$, $\dfrac{1+1}{2+1} = \dfrac{2}{3}$, $\dfrac{2+1}{3+1} = \dfrac{3}{4}$, $\dfrac{3+1}{4+1} = \dfrac{4}{5}$, $\dfrac{4+1}{5+1} = \dfrac{5}{6}$

None of the resulting fractions, 1/2, 2/3, 3/4, 4/5, and 5/6, **are
equivalent**. This is because the relative sizes of the numerators
are increasing faster than the denominators. For example, adding
1 to the numerator of 1/2 doubles it size (e.g. 1 + 1 = 1 x 2). To
double the denominator of 1/2, you have to add not 1 but 2 (e.g.
2 + 2 = 2 x 2). The idea is that you need to preserve the ratio of
the top and bottom numbers for the fractions to be equivalent.

(e) The total number of eggs is 36 + 48 = 84. So **your fraction
of the eggs is 36/84**. Next, your fraction of the cartons: You
can fill 3 cartons (36 ÷ 12 = 3). Your friend can fill 4 cartons
(48 ÷ 12 = 4). The total number of cartons is 3 + 4 = 7. So **your
fraction of the cartons is 3/7. To show your fractions are
equivalent, or 3/7 = 36/84, you can multiply the numerator
and denominator of 3/7 by 12**, which is the number of eggs per
carton. **This keeps them the same relative size.** It really is
like multiplying by 12/12, which is the same as 1.

$$\frac{3}{7} = \frac{3 \times 12}{7 \times 12} = \frac{36}{84}$$

(f) The fraction, $\dfrac{\text{Sam's age}}{\text{Grandmother's age}} = \dfrac{?}{72} = \dfrac{1}{8}$

Using cross multiplication: $(72)(1) = (8)(?)$

What is 8 multiplied by to equal 72? Using the multiplication table, see that: $8 \times 9 = 72$, so **Sam is now 9 years old.**

In 9 more years, Sam will be $9 + 9 = 18$ and grandmother will be $72 + 9 = 81$. Then Sam will be **18/81 of grandmother's age, which equals 2/9.** (We divided top and bottom by 9.) Note that $1/8 = 9/72$ and $2/9 = 16/72$, so $2/9$ is greater than $1/8$.

(g) Using equivalent fractions: $\dfrac{3}{40} = \dfrac{15}{?}$

Since $3 \times 5 = 15$, then $\dfrac{3 \times 5}{40 \times 5} = \dfrac{15}{200}$

So, his collection has 200 marbles.

4.2

(a) We can add and subtract the dimes or use tenths of a dollar.
Using dimes: $5 + 6 + 6 - 2 - 2 - 2 = $ **11 dimes left in the jar.**

Using tenths of a dollar:

$5/10 + 6/10 + 6/10 - 2/10 - 2/10 - 2/10 = ?$

adding: $\dfrac{5 + 6 + 6 - 2 - 2 - 2}{10} = \dfrac{11}{10}$ of a dollar left in the jar

Also, $\dfrac{10}{10} + \dfrac{1}{10} = 1 + \dfrac{1}{10}$ or one dollar and ten cents left in the jar

(b) 30 people each want 3 slices, which is 30 x 3 = 90 slices needed. You have 9 slices already. So you need to make 90 - 9 = 81 slices. So you need 81 slices to add to the 9 slices you already have. But how many pizzas do you need to make? Because each pizza is cut in 8ths, we divide 81 by 8 to see how many pizzas we need. 8 goes into 81 ten times with 1 left over. Remember 8 x 10 = 80, so 81/8 = 10 plus 1/8 pizzas. **You need to make 11 pizzas total**, but you only need 1 piece from the 11th pizza. **You will have 7 pieces left**.

Let's think about how we solved this problem. Because all the denominators are the same (they are 1/8ths of a pizza) we have solved this problem using the numerators. Then we divided by 8 to get the number of pizzas. We can also work this problem using fractions. Let's look at that:

If 30 people each get 1 slice of pizza: 30 x 1/8 = 30/8

If 30 people each get 3 slices of pizza: $30 \times 3/8 = \dfrac{30 \times 3}{8}$

$$= \frac{30}{8} \times 3 = \frac{30}{8} + \frac{30}{8} + \frac{30}{8} = \frac{30 + 30 + 30}{8} = \frac{90}{8} \text{ pizzas}$$

The restaurant already has 9 slices, so it needs to bake

$$\frac{90}{8} - \frac{9}{8} = \frac{81}{8} \text{ pizzas}$$

How many 8-slice pizzas does it take to give you 81 slices? 81/8 = 10 r 1. So, to get all the slices it needs, the restaurant **needs to bake 11 more pizzas** and take the 1 extra slice from the 11th pizza. If you take 1 slice from the 11th pizza you will have 8/8 – 1/8 = 7/8 of a **pizza left, or 7 slices**.

4.3

(a) These are all fractions of a dollar: a penny is 1/100 of a dollar, a nickel is 1/20, a dime is 1/10, a quarter is 1/4, and a half dollar is 1/2. To add them you must find a common denominator. They all can be expressed in cents, which are hundredths of a dollar. So the calculation is:

1/100 + 5/100 + 10/100 + 25/100 + 50/100

$$= \frac{1 + 5 + 10 + 25 + 50}{100} = \frac{91}{100} \text{ or } \textbf{91 cents}$$

We can also convert to cents add:
1 cent + 5 cents + 10 cents + 25 cents + 50 cents = **91 cents**

(b) You ate: 1/6 + 1/8 + 1/10 + 1/12 = ? of a pizza. Now add them:

The first addition is 1/6 + 1/8, or

$$\frac{1}{6} \ast \frac{1}{8} = \frac{(8 \times 1) + (6 \times 1)}{6 \times 8} = \frac{8 + 6}{48} = \frac{14}{48} \text{ Divide } \frac{\div 2}{\div 2} = \frac{7}{24}$$

The second addition is 7/24 + 1/10, or $\dfrac{7}{24} \ast \dfrac{1}{10}$

$$= \frac{(10 \times 7) + (24 \times 1)}{24 \times 10} = \frac{70 + 24}{240} = \frac{94}{240} \text{ Divide } \frac{\div 2}{\div 2} = \frac{47}{120}$$

The last addition is 47/120 + 1/12, or $\dfrac{47}{120} \ast \dfrac{1}{12}$

$$= \frac{(12 \times 47) + (120 \times 1)}{120 \times 12} = \frac{564 + 120}{1440} = \frac{684}{1440} \text{ Now reduce:}$$

Let's reduce using divisions by prime numbers 2, 3, 5, etc. Start with 2 until it no longer divides exactly, then try 3, etc.:

$$\frac{684 \div 2}{1440 \div 2} = \frac{342}{720} \quad again \quad \frac{\div 2}{\div 2} = \frac{171}{360}$$

$$next \quad \frac{\div 3}{\div 3} = \frac{57}{120} \quad again \quad \frac{\div 3}{\div 3} = \frac{19}{40}$$

$\frac{19}{40}$ does not reduce further and is just under one half.

So, you ate just under half a pizza.

Note: An alternative method for solving this type of problem is to first find the "lowest common denominator". This is described in Master Math: Basic Math and Pre-Algebra Section 2.4.

(c) You first add the 7/8 ton with the 1/3 ton:

$$\frac{7}{8} \times \frac{1}{3} = \frac{(3 \times 7) + (8 \times 1)}{24} = \frac{21 + 8}{24} = \frac{29}{24}$$

Then you subtract the 2/5 ton used for your garden to see how much is left:

$$\frac{29}{24} \times \frac{2}{5} = \frac{(29 \times 5) - (24 \times 2)}{120} = \frac{145 - 48}{120} = \frac{97}{120}$$

So you have **97/120** tons left to spread on your lawn. If you divide 97 by 120 you get about 0.8 tons, or about 8/10 of a ton.

Chapter 5
Multiplying & Dividing Simple and Complex Fractions

... ye shall seek me, and find me, when ye shall search for me with all your heart. Jer29:13

Here's a preview of multiplying and dividing fractions

To multiply simple fractions just multiply their numerators with numerators and denominators with denominators:

$$\frac{\text{numerator 1}}{\text{denominator 1}} \times \frac{\text{numerator 2}}{\text{denominator 2}} = \frac{\text{numerator 1} \times \text{numerator 2}}{\text{denominator 1} \times \text{denominator 2}}$$

To divide simple fractions flip the 2nd fraction upside-down and multiply the numerators and denominators:

$$\frac{\text{numerator 1}}{\text{denominator 1}} \div \frac{\text{numerator 2}}{\text{denominator 2}} = \frac{\text{numerator 1}}{\text{denominator 1}} \times \frac{\text{denominator 2}}{\text{numerator 2}}$$

$$= \frac{\text{numerator 1} \times \text{denominator 2}}{\text{denominator 1} \times \text{numerator 2}}$$

But before we learn more about multiplication and division let's...

5.1. Reduce Fractions

Why would I reduce a fraction and how would I do it?
Reducing fractions make them easier to work with. And to
reduce fractions you either divide both the numerator and
denominator by the same number or you cancel common factors.

Before showing how to reduce a fraction let's, use pizzas to
LOOK at how reducing fractions works! We can think of
"unslicing" a pizza that has been sliced into quarters or sixths.

**Example: If you have 2/4 of a pizza left over, how much of
the pizza really remains.**

          Unslice

  2/4 pizza   Same as 1/2 pizza

Unslicing it shows there is really 1/2 left.

We can write this: $\dfrac{2 \div 2}{4 \div 2} = \dfrac{1}{2}$

**Example: You have 4/6 of a pizza left over. How much of the
pizza really remains.**

          Unslice

  4/6 pizza   Shows 2/3 pizza

Unslicing it shows there is really 2/3 left.

We can write this: $\dfrac{4 \div 2}{6 \div 2} = \dfrac{2}{3}$

Here are two methods you can use to reduce fractions

Method 1: Divide both the numerator and denominator by a number that divides exactly into each of them. If possible, continue to divide small numbers into the numerator and denominator. Or you may identify a larger number that divides exactly into both the numerator and denominator. For example:

$$\frac{8 \div 2}{12 \div 2} = \frac{4}{6} \quad \text{then} \quad \frac{4 \div 2}{6 \div 2} = \frac{2}{3} \qquad \text{Or} \qquad \frac{8 \div 4}{12 \div 4} = \frac{2}{3}$$

Method 2: First write both the numerator and denominator in their factored forms. Then cancel factors that are present in both the numerator and denominator.
For example:

$$\frac{8}{12} = \frac{2 \times 2 \times 2}{2 \times 2 \times 3} = \frac{\cancel{2} \times \cancel{2} \times 2}{\cancel{2} \times \cancel{2} \times 3} = \frac{2}{3}$$

This works because $\frac{2}{2}$ = 1, so you're not changing the value!

Remembering **FACTORS**: We learned in Section 3.3 that a **factor is a smaller number that can be divided exactly into a larger number** without a remainder. For example, both 2 and 3 are factors of 6 since 6 ÷ 3 = 2 and 6 ÷ 2 = 3. Also, 2 x 3 = 6.

Numbers that are multiplied together to produce a product are factors of that product. You find the factors by seeing what divides exactly into your number.

When you are solving problems that have fractions, you can sometimes reduce a fraction before you solve a problem. After you finish adding, subtracting, multiplying, or dividing fractions, you may also need to reduce your answer.

Let's practice reducing!

Example: Reduce 2/16.

Since 2 divides exactly into the 2 and the 16 in fraction 2/16, we can reduce 2/16 by dividing its top and bottom numbers by 2. Let's see:

$$\frac{2 \div 2}{16 \div 2} = \frac{1}{8} \qquad \textbf{Therefore, 2/16 = 1/8}$$

Example: Reduce 24/60.

Let's try **factoring** top and bottom:

$$\frac{24}{60} = \frac{2 \times 2 \times 2 \times 3}{2 \times 2 \times 3 \times 5} = \frac{\cancel{2} \times \cancel{2} \times 2 \times \cancel{3}}{\cancel{2} \times \cancel{2} \times \cancel{3} \times 5} = \frac{2}{5}$$

Or, we could have noticed that 12 divides exactly into 24 and 60:

$$\frac{24 \div 12}{60 \div 12} = \frac{2}{5}$$

Therefore, **24/60 = 2/5**.

Example: Reduce 25/40.

Let's try factoring top and bottom:

$$\frac{25}{40} = \frac{5 \times 5}{\underbrace{5 \times 2}_{10} \times \underbrace{2 \times 2}_{4}} = \frac{\cancel{5} \times 5}{\cancel{5} \times 2 \times 2 \times 2} = \frac{5}{8}$$

Or, we could have noticed that 5 divides exactly into 25 and 40:

$$\frac{25 \div 5}{40 \div 5} = \frac{5}{8}$$

Therefore, 25/40 = 5/8.

Example: Reduce 21/40.

Let's try factoring top and bottom:

$$\frac{21}{40} = \frac{3 \times 7}{\underbrace{5 \times 2}_{10} \times \underbrace{2 \times 2}_{4}} = \frac{3 \times 7}{5 \times 2 \times 2 \times 2} = \frac{21}{40}$$

There are no factors that occur in both the numerator and the denominator. Another way to say this is: there are no factors "common to" both the numerator and denominator.
Since no number divides exactly into both the numerator 21 and denominator 40, you cannot reduce 21/40.
Therefore, 21/40 cannot be reduced.

5.2. Multiply Fractions

$$\frac{\text{numerator 1}}{\text{denominator 1}} \times \frac{\text{numerator 2}}{\text{denominator 2}} = \frac{\text{numerator 1} \times \text{numerator 2}}{\text{denominator 1} \times \text{denominator 2}}$$

To multiply fractions:

1. Multiply the numerators.

2. Multiply the denominators.

3. Place product of numerators over product of denominators.

You may reduce answer by dividing top and bottom by common factors.

Example: Multiply 2/3 by 5/8.

We multiply the numerators and multiply the denominators:

$$\frac{2}{3} \times \frac{5}{8} = \frac{2 \times 5}{3 \times 8} = \frac{10}{24} = \frac{5}{12}$$

Note: We reduced 10/24 by dividing both 10 and 24 by 2.

Or first reduce by canceling: $\dfrac{2 \times 5}{3 \times 8} = \dfrac{2 \times 5}{3 \times 2 \times 4} = \dfrac{1 \times 5}{3 \times 1 \times 4} = \dfrac{5}{12}$

Therefore, 2/3 x 5/8 = 5/12.

This is the easiest way to multiply fractions.

Let's Understand What Multiplication of Fractions Means

When you multiply a fraction by a fraction,   1/2 x 1/2

you **take a fraction of the first fraction**:   take 1/2 of 1/2

On a number line you can see 1/2 x 1/2 as

taking a fraction of the first fraction, or

taking 1/2 of 1/2 which is 1/4

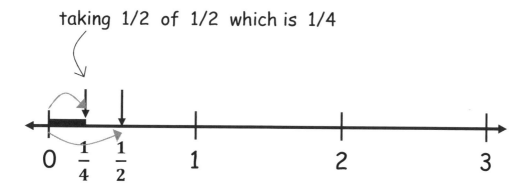

Therefore, 1/2 x 1/2 = 1/4

Next we'll use pizzas to LOOK at multiplying 1/2 x 1/2.
If you have **a half of a pizza times one-half,** you have
a half of a half of a pizza. Let's see:

Example: Multiply 1/2 x 1/2.

          x 1/2   =

1/2 pizza   take 1/2 of the 1/2   leaves 1/4 pizza

We can also write it: $\dfrac{1}{2} \overset{\longleftrightarrow}{\times} \dfrac{1}{2} = \dfrac{1 \times 1}{2 \times 2} = \dfrac{1}{4}$

Example: Show multiplying 1/3 x 1/2.

          x 1/2   =

1/3 pizza   take 1/2 of the 1/3   leaves 1/6 pizza

We can also write it: $\dfrac{1}{3} \overset{\longleftrightarrow}{\times} \dfrac{1}{2} = \dfrac{1 \times 1}{3 \times 2} = \dfrac{1}{6}$

Example: What is 1/2 x 1/3? This means find 1/3 of 1/2?

          x 1/3   =

1/2 pizza   take 1/3 of the 1/2   leaves 1/6 pizza

We can also write it: $\dfrac{1}{2} \overset{\longleftrightarrow}{\times} \dfrac{1}{3} = \dfrac{1 \times 1}{2 \times 3} = \dfrac{1}{6}$

Remember, multiplication is commutative: order doesn't matter.

Example: A piece of land is 3/5 miles long and 7/8 miles wide. How many square miles is it? (Area = length x width.)

Just multiply the numerators and multiply the denominators:

$$\frac{3}{5} \times \frac{7}{8} = \frac{3 \times 7}{5 \times 8} = \frac{21}{40} \text{ square miles}$$

3/5 mi

7/8 mi

Therefore, the land is 21/40 or 0.525 square miles. (We can't reduce 21/40 since nothing divides exactly into both 21 and 40.)

If a fraction's top number is smaller than its bottom number, its value is less than 1. If you multiply two fractions whose top numbers are smaller than their bottom numbers, then the value of the product will be less than the value of either multiplying fraction. This is true because multiplying takes a fraction of a fraction. For example, using the above:
3/5 = 0.6, 7/8 = 0.875, and 21/40 = 0.525

Multiplying More Than 2 Fractions

What about multiplying more than 2 fractions?

1. Multiply all the numerators together.

2. Multiply all the denominators together.

3. Place product of numerators over product of denominators.

May reduce answer by dividing top and bottom by common factors.

Example: Multiply 2/5 x 5/6 x 7/9.

We multiply the numerators and multiply the denominators:

$$\frac{2}{5} \times \frac{5}{6} \times \frac{7}{9} = \frac{2 \times 5 \times 7}{5 \times 6 \times 9} = \frac{70}{270} = \frac{7}{27}$$

Note: we reduced 70/270 by dividing both 70 and 270 by 10.

Or reduce first: $\dfrac{2 \times 5 \times 7}{5 \times 6 \times 9} = \dfrac{2 \times 5 \times 7}{5 \times 2 \times 3 \times 9} = \dfrac{1 \times 7}{1 \times 3 \times 9} = \dfrac{7}{27}$

Therefore, 2/5 x 5/6 x 7/9 = 7/27.

Let's LOOK at multiplying more than 2 fractions using pizza.

Example: Multiply 1/2 x 1/2 x 1/2.

You can draw taking **a half** of **a half** of **a half** of a pizza:

1/2 pizza   take 1/2   leaves 1/4   take 1/2   leaves 1/8
                             pizza   pizza

You can also write it: $\dfrac{1}{2} \times \dfrac{1}{2} \times \dfrac{1}{2} = \dfrac{1 \times 1 \times 1}{2 \times 2 \times 2} = \dfrac{1}{8}$

Example: Multiply (12/15)(12/14)(12/9)(12/20)(2).

To multiply a bunch of simple fractions we just multiply the numerators and multiply the denominators. But first notice there are a couple of fractions that we can reduce first. Reducing makes the fractions easier to work with. Let's look at each fraction and reduce it by dividing by common factors.

$$\frac{12 \div 3}{15 \div 3} = \frac{4}{5} \qquad \frac{12 \div 2}{14 \div 2} = \frac{6}{7} \qquad \frac{12 \div 3}{9 \div 3} = \frac{4}{3}$$

$$\frac{12 \div 4}{20 \div 4} = \frac{3}{5} \qquad \text{and also note that } 2 = \frac{2}{1}$$

Now we can multiply 12/15 x 12/14 x 12/9 x 12/20 x 2 as:

$$\frac{4}{5} \times \frac{6}{7} \times \frac{4}{3} \times \frac{3}{5} \times \frac{2}{1} = \frac{4 \times 6 \times 4 \times \cancel{3} \times 2}{5 \times 7 \times \cancel{3} \times 5 \times 1}$$

Cancel common 3s

$$= \frac{4 \times 6 \times 4 \times 2}{5 \times 7 \times 5 \times 1} = \frac{192}{175}$$

Therefore, 12/15 x 12/14 x 12/9 x 12/20 x 2 = 192/175.

5.3. Divide Fractions

$$\frac{\text{numerator 1}}{\text{denominator 1}} \div \frac{\text{numerator 2}}{\text{denominator 2}} = \frac{\text{numerator 1}}{\text{denominator 1}} \times \frac{\text{denominator 2}}{\text{numerator 2}}$$

To divide fractions:

1. Flip the 2nd fraction upside-down (called a <u>reciprocal</u>).

2. Then multiply fractions as usual by
   multiplying the (new) numerators and the (new) denominators.

If needed, we can then reduce the answer by dividing top and bottom by common factors or canceling common factors before dividing.

Let's define reciprocal fraction: A reciprocal fraction is a fraction turned upside down.

Dividing by a fraction is the same as multiplying by the reciprocal.

Here are examples of reciprocals:

$$\frac{1}{2} \xrightarrow{\text{flip}} \frac{2}{1} \qquad \frac{2}{3} \xrightarrow{\text{flip}} \frac{3}{2} \qquad \frac{5}{2} \xrightarrow{\text{flip}} \frac{2}{5}$$

fraction reciprocal   fraction reciprocal   fraction reciprocal

If you multiply a fraction by its reciprocal you get 1.

Let's try 2/3:

$$\frac{2}{3} \times \frac{3}{2} = \frac{2 \times 3}{3 \times 2} = \frac{6}{6} = 1$$

Or we can cancel common factors: $\dfrac{2}{3} \times \dfrac{3}{2} = \dfrac{2 \times 3}{3 \times 2} = 1$

Let's try an example where we divide fractions:

Example: Divide 1/2 by 1/2.

Flip the 2nd fraction upside-down and multiply the fractions:

$$\frac{1}{2} \div \frac{1}{2} = \frac{1}{2} \times \frac{2}{1} = \frac{1 \times 2}{2 \times 1} = \frac{2}{2} = 1$$

This is the easiest way to divide fractions.

Before we show more examples, let's look at dividing fractions on a number line.

We Can See Dividing Fractions on a Number Line

Example: On a number line show 1/2 ÷ 1/2.

Begin at 1/2. See how many 1/2s fit into 1/2:

<u>ONE</u> **1/2** can fit into (or be subtracted from) 1/2 before reaching 0.

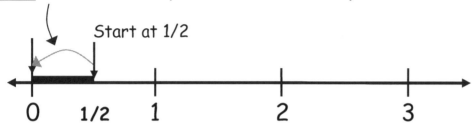

Therefore, 1/2 ÷ 1/2 = 1

Let's Look Further at the Division of Fractions

Division calculates how many times one number is present in, or "fits" into, another number.

What does 1/2 ÷ 1/2 really say?

> 1/2 ÷ 1/2 says, "How many 1/2 's fit into 1/2?"

We can see that exactly **ONE** 1/2 will fit into 1/2:

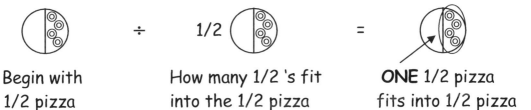

| Begin with | How many 1/2 's fit | **ONE** 1/2 pizza |
| 1/2 pizza | into the 1/2 pizza | fits into 1/2 pizza |

1/2 ÷ 1/2 = 1 since ONE 1/2 pizza fits into a half pizza.

Example: Show and calculate 1/2 ÷ 1/4.

Remember, 1/2 ÷ 1/4 asks, "How many 1/4 's fit into 1/2?"

Seems we could fit TWO 1/4 pizzas into 1/2 pizza, but let's look:

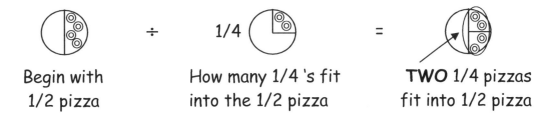

| Begin with | How many 1/4 's fit | **TWO** 1/4 pizzas |
| 1/2 pizza | into the 1/2 pizza | fit into 1/2 pizza |

Now calculate: Flip the 2nd fraction upside-down and multiply.

$$\frac{1}{2} \div \frac{1}{4} = \frac{1}{2} \times \frac{4}{1} = \frac{1 \times 4}{2 \times 1} = \frac{4}{2} = 2 \quad \text{So, } 1/2 \div 1/4 = 2$$

Therefore, 1/2 ÷ 1/4 = 2 or TWO 1/4 pizzas fit into 1/2 pizza.

Let's write division as a fraction to show that flipping works!

Look at 2/3 ÷ 4/5:

$$\frac{2}{3} \div \frac{4}{5} \qquad \text{Write} \div \text{as a fraction:} \qquad \frac{2/3}{4/5}$$

First multiply by 3/3 to eliminate the 3 in the numerator:

$$\frac{2/3}{4/5}\frac{\times 3}{\times 3} = \frac{\left(\frac{2}{3}\right)(3)}{\left(\frac{4}{5}\right)(3)} = \frac{\left(\frac{2}{\cancel{3}}\right)\cancel{(3)}}{\left(\frac{4}{5}\right)(3)} = \frac{2}{\left(\frac{4}{5}\right)(3)}$$

Then multiply by 5/5 to eliminate the 5 in the denominator:

$$= \frac{2}{\left(\frac{4}{5}\right)(3)}\frac{\times 5}{\times 5} = \frac{2\times 5}{(3)\left(\frac{4}{\cancel{5}}\right)\cancel{\times 5}} = \frac{2\times 5}{3\times 4} = \frac{2}{3} \times \frac{5}{4}$$

We see it works out that $\quad \dfrac{2}{3} \div \dfrac{4}{5} \; = \; \dfrac{2}{3} \times \dfrac{5}{4}$

Note that this equals $\quad \dfrac{10}{12} = \dfrac{5}{6}$

Dividing More Than 2 Fractions

What about dividing 3 fractions? To divide 3 fractions:

1. Flip the 2nd and 3rd fraction upside-down.

2. Then multiply fractions as usual by
   multiplying the (new) numerators and the (new) denominators.

Note that this gives the same answer as dividing in order using:
Divide first 2 fractions by "flip 2nd fraction and multiply".
Then divide answer by 3rd fraction using "flip and multiply".

You may reduce the answer or cancel common factors in the numerators and denominators before multiplying. You may not need to factor all factorable numbers if you see nothing more can be canceled.

Example: Divide 3/8 ÷ 6/7 ÷ 9/10.

Flip the 2nd and 3rd fractions upside-down and then multiply:

$$\frac{3}{8} \div \frac{6}{7} \div \frac{9}{10} = \frac{3}{8} \times \frac{7}{6} \times \frac{10}{9} = \frac{3 \times 7 \times 10}{8 \times 6 \times 9} = \frac{3 \times 7 \times 2 \times 5}{8 \times 3 \times 2 \times 9} = \frac{35}{72}$$

Do we get the same answer as dividing in order? Let's see:
First divide 3/8 ÷ 6/7 using "flip and multiply":

$$\frac{3}{8} \div \frac{6}{7} = \frac{3}{8} \times \frac{7}{6} = \frac{3 \times 7}{8 \times 6} = \frac{3 \times 7}{8 \times 3 \times 2} = \frac{7}{16}$$

Next divide 7/16 ÷ 9/10 using "flip and multiply":

$$\frac{7}{16} \div \frac{9}{10} = \frac{7}{16} \times \frac{10}{9} = \frac{7 \times 10}{16 \times 9} = \frac{7 \times 2 \times 5}{8 \times 2 \times 9} = \frac{35}{72}$$

Therefore, 3/8 ÷ 6/7 ÷ 9/10 = 35/72.

Combinations of Multiplying and Dividing

To find the answer when you have more than 2 fractions that are multiplied and divided with each other:

1. Multiply numerators and denominators of multiplied fractions.

2. Flip divided fractions, then multiply flipped fractions by multiplying their (new) numerators and (new) denominators.

You may reduce before **or** after multiplying by dividing top and bottom by common factors.

For example:

$$\frac{1}{2} \times \frac{1}{2} \div \frac{1}{2} \times \frac{1}{2} \div \frac{1}{2} = \frac{1}{2} \times \frac{1}{2} \times \frac{2}{1} \times \frac{1}{2} \times \frac{2}{1} = \frac{4}{8} = \frac{1}{2}$$

You can also cancel before multiplying: $= \dfrac{1 \times 1 \times \cancel{2} \times 1 \times \cancel{2}}{2 \times \cancel{2} \times 1 \times \cancel{2} \times 1} = \dfrac{1}{2}$

Example: Solve 3/5 ÷ 5/6 × 13/14.

Flip the 2nd fraction upside-down and multiply all fractions:

$$\frac{3}{5} \div \frac{5}{6} \times \frac{13}{14} = \frac{3}{5} \times \frac{6}{5} \times \frac{13}{14} = \frac{3 \times 6 \times 13}{5 \times 5 \times 14} = \frac{3 \times \cancel{2} \times 3 \times 13}{5 \times 5 \times \cancel{2} \times 7} = \frac{117}{175}$$

Therefore, **3/5 ÷ 5/6 × 13/14 = 117/175.** = .6685

Example: Your truck can carry a load of 3/4 tons. You make 4 trips to the garden store to buy patio blocks weighing 1/60 ton each and fully load your truck each time. How many patio blocks did you move?

Let's think this through using given information and units. To find total blocks moved we need to know how many blocks fit into a full truckload. We know there are 4 truckloads. We write:

Total blocks = blocks per full truckload x 4 truckloads

 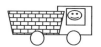

How many blocks per full truckload? We are told:
The truck capacity is 3/4 ton and each block weighs 1/60 ton.
In division, the (numerator/denominator) means how many denominators fit into the numerator.
If you divide the total weight of a truckload, 3/4 ton, by the weight of a single block, 1/60 ton, you get the number of blocks.

Weight of each block 1/60 ton

Maximum weight of a truckload is 3/4 ton

$$\text{Number of blocks in truckload} = \frac{\text{truckload weight}}{\text{weight per block}} = \frac{\frac{3}{4}\,\text{ton}}{\frac{1}{60}\,\text{ton/block}}$$

So, the number of blocks in a truckload = 3/4 ton ÷ 1/60 ton/block

Now we need total blocks in 4 truckloads. We can write this as:

Total blocks = number of blocks in a truckload × 4 truckloads, **or**

Total blocks = number of blocks per truckload × 4 truckloads

Putting it all together our equation is:

Total blocks = 3/4 ton/truckload ÷ 1/60 ton/block × 4 truckloads

$$\text{Total blocks} = \frac{3}{4}\frac{\text{ton}}{\text{truckload}} \div \frac{1}{60}\frac{\text{ton}}{\text{blocks}} \times \frac{4\text{ truckloads}}{1}$$

$$\text{Total blocks} = \frac{3}{4}\frac{\text{ton}}{\text{truckload}} \times \frac{60}{1}\frac{\text{blocks}}{\text{ton}} \times \frac{4\text{ truckloads}}{1}$$

Cancel like units and numbers:

$$\text{Total blocks} = \frac{3}{4}\frac{\cancel{\text{ton}}}{\cancel{\text{truckload}}} \times \frac{60\text{ blocks}}{1\cancel{\text{ton}}} \times \frac{\cancel{4\text{ truckloads}}}{1}$$

Total blocks = 3 × 60 blocks = 180 blocks

Therefore, the total number of blocks is 180.

SEE how easy it was to solve this complicated problem?
Now, you can see the importance of learning fractions!

Dividing a Combination of Fractions and Integers

Sometimes you will see a division problem that includes both fractions (e.g. 2/3) and integers (e.g. 8). It is often helpful to turn the integer into a fraction by making the integer the numerator and placing a 1 as its denominator: 8 = 8/1
Then proceed with the division by flipping the divided fraction upside down and multiplying it to the other fraction. For example:

$$8 \div 2/3 \ = \ \frac{8}{1} \div \frac{2}{3} \ = \ \frac{8}{1} \times \frac{3}{2} \ = \ \frac{24}{2} \ = \ 12$$

In summary, to divide fractions with integers?

1. Divide the integers by 1, such as 3 = 3/1.

2. Flip the divided fractions upside-down and multiply.

Example divide 3 ÷ 5/6 and 2/3 ÷ 5.

First see that 3 = 3/1 and 5 = 5/1. Now solve:

$$\frac{3}{1} \div \frac{5}{6} \ = \ \frac{3}{1} \times \frac{6}{5} \ = \ \frac{3 \times 6}{1 \times 5} \ = \ \frac{18}{5}$$

$$\frac{2}{3} \div \frac{5}{1} \ = \ \frac{2}{3} \times \frac{1}{5} \ = \ \frac{2 \times 1}{3 \times 5} \ = \ \frac{2}{15}$$

Therefore, **3 ÷ 5/6 = 18/5** and **2/3 ÷ 5 = 2/15**.

5.4. Mixed Numbers and Improper Fractions

Improper Fractions

Improper fractions are just fractions with the top number bigger than the bottom number.

$$\text{Improper Fraction} = \frac{\textbf{Bigger Top Number}}{\textbf{Smaller Bottom Number}}$$

$\dfrac{7}{6}$, $\dfrac{15}{7}$,   and $\dfrac{238}{107}$ are all **improper fractions**.

We can use pizza to SEE what improper fractions look like:

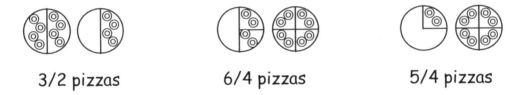

| 3/2 pizzas | 6/4 pizzas | 5/4 pizzas |

All improper fractions have a value greater than 1.

You can write every integer as an improper fraction by giving it a denominator of 1: (integer/1)

Example: Write 5, 22, and 101 as improper fractions.

$$5 = \frac{5}{1}, \quad 22 = \frac{22}{1}, \quad \text{and} \quad 101 = \frac{101}{1}$$

You add, subtract, multiply, and divide improper fractions just as you do simple fractions.

Mixed Numbers

If an integer and a simple fraction are written together to describe a value, it is called a mixed number. Let's see:

$$\text{Mixed Number} = \text{Integer}\frac{\text{Numerator}}{\text{Denominator}} = \text{Integer} + \frac{\text{Numerator}}{\text{Denominator}}$$

$$3\frac{5}{6}, \quad 1\frac{1}{6}, \quad 2\frac{1}{7}, \quad \text{and} \quad 2\frac{24}{107} \quad \text{are mixed numbers}$$

A mixed number shows the whole plus the fraction left over. Note: the number 3 5/6 is pronounced "three and five-sixths".

Can We See Mixed Numbers on the Number Line? Yes!

Example: Show mixed number 2 1/2 on the number line:

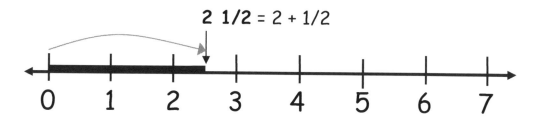

Example: Show mixed number 3 3/4 on the number line:

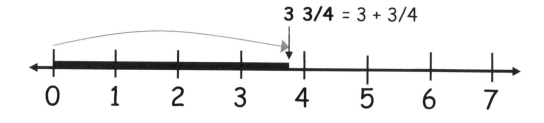

Example: Show mixed number -2 1/2 on the number line:

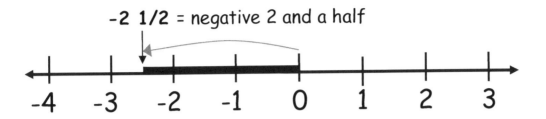

-2 1/2 = negative 2 and a half

Mixed numbers can be converted to improper fractions.
Improper fractions can be converted to mixed numbers.

Mixed Numbers      Improper Fractions

Mixed numbers are improper fractions written differently:

3/2 pizzas

1 1/2 pizzas

6/4 pizzas

1 2/4 pizzas

5/4 pizzas

1 1/4 pizzas

See that:

3/2 = 1 1/2 = 1 + 1/2   6/4 = 1 2/4 = 1 + 2/4   5/4 = 1 1/4 = 1 + 1/4

Improper fractions (e.g. 3/2) are usually easier to work with than mixed numbers (e.g. 1 1/2) when performing calculations!

How Do You Convert Improper Fractions into Mixed Numbers

To convert an improper fraction into a mixed number:

   Divide the numerator by the denominator.

$$\frac{\text{Improper fraction's (larger) Numerator}}{\text{Improper fraction's (smaller) Denominator}} = \text{Mixed Number}$$

In other words,

| Improper fraction's Numerator | ÷ | Improper fraction's Denominator | = | Mixed Number |

Example: Write improper fractions $\frac{7}{6}$ and $\frac{17}{7}$ as mixed numbers.

To convert an improper fraction into a mixed number, just divide:

$7 \div 6 = ?$ We know 6 divides into 7 ONE time with 1/6 left over.

$17 \div 7 = ?$ We know 7 divides into 17 TWO times with 3/7 left over.

We can see these using long division:

$$6\overline{)7} \quad\quad 6\overline{)7}\;\;1/6 \quad\quad 7\overline{)17} \quad\quad 7\overline{)17}\;\;3/7$$

with the worked divisions showing:

$6\overline{)7}$ with quotient 1, -6, remainder 1

$7\overline{)17}$ with quotient 2, -14, remainder 3

Therefore: $\frac{7}{6} = 1\frac{1}{6}$ and $\frac{17}{7} = 2\frac{3}{7}$

How Do You Convert Mixed Numbers into Improper Fractions

One way to convert a mixed number into an improper fraction:

1. Multiply the denominator by the integer.

2. Add product to the numerator to get a new numerator.

3. Place the new numerator over the denominator.

Let's show this:

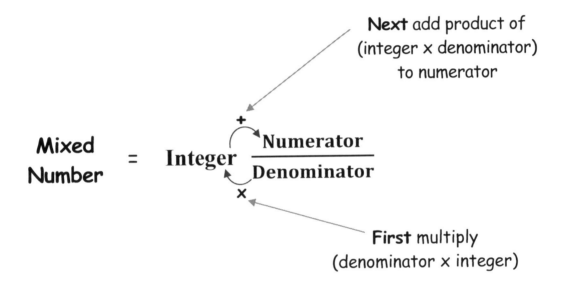

$$= \frac{(\text{Denominator} \times \text{Integer}) + \text{Numerator}}{\text{Denominator}} = \text{Improper Fraction}$$

Example: Write 1 1/6 and 2 3/7 as improper fractions.

First, multiply the denominator by the integer.
Second, add the product to numerator to get a new numerator.
Finally, put new numerator over denominator.

2nd, +

$$1\,\frac{1}{6} = \frac{(6 \times 1) + 1}{6} = \frac{6 + 1}{6} = \frac{7}{6} \qquad \text{So, } 1\,\frac{1}{6} = \frac{7}{6}$$

1st, ✗

2nd, +

$$2\,\frac{3}{7} = \frac{(7 \times 2) + 3}{7} = \frac{14 + 3}{7} = \frac{17}{7} \qquad \text{So, } 2\,\frac{3}{7} = \frac{17}{7}$$

1st, ✗

An additional way to convert a mixed number into an improper fraction is to add the integer and its fraction:

1. Make the integer a fraction by putting it over 1.
2. Find a common denominator.
3. Add the two fractions to get the improper fraction.

For example, if we have 2 1/4, we can write: $2\,\frac{1}{4} = \frac{2}{1} + \frac{1}{4}$

Then we find a common denominator and add the fractions.

Here are the steps to adding the integer and its fraction.

If you have **mixed number**: $2\dfrac{1}{4}$

Write it as **integer + fraction**, then set the integer over 1:

$$2 + \frac{1}{4} \;=\; \frac{2}{1} + \frac{1}{4}$$

Find a **common denominator** and add:

$$\frac{2}{1} \times\!\!+\!\!\times \frac{1}{4} \;=\; \frac{(2 \times 4) + (1 \times 1)}{1 \times 4}$$

$$=\; \frac{8 + 1}{4} \;=\; \frac{9}{4} \;,\; \text{which is the \textbf{improper fraction}}$$

We can see this with pizzas:

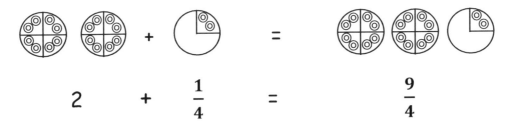

$$2 \qquad + \qquad \frac{1}{4} \qquad = \qquad \frac{9}{4}$$

The reason we can add the integer and its fraction is that a **mixed number** is just **the whole plus the fraction left over**.

Example: Write 1 1/2 as an improper fraction by adding the integer and its fraction.

Write the mixed number as integer plus fraction.
Then write the integer as a fraction by putting it over 1

$1\ 1/2\ =\ 1 + 1/2\ =\ 1/1 + 1/2$

Find common denominator and add to get the improper fraction:

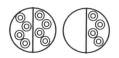 $\dfrac{1}{1} \ast \dfrac{1}{2}\ =\ \dfrac{(1 \times 2) + (1 \times 1)}{1 \times 2}\ =\ \dfrac{2 + 1}{2}\ =\ \dfrac{3}{2}$ **So, 1 1/2 = 3/2**

We see: 1 1/2 = 3/2

In summary: When you convert a mixed number into an improper fraction, it is usually faster to use the first method we showed:
Multiply (denominator x integer);
Add product to the numerator to get a new numerator;
Put the new numerator over the denominator.

Next add (denominator x integer) to numerator

$$\text{Mixed Number}\ =\ \textbf{Integer}\ \dfrac{\textbf{Numerator}}{\textbf{Denominator}}\ =\ \begin{array}{c}\text{Improper}\\ \text{Fraction}\end{array}$$

First multiply (denominator x integer)

5.5. Add, Subtract, Multiply, and Divide Mixed Numbers

Adding and Subtracting Mixed Numbers

To add or subtract mixed numbers:

First convert the mixed numbers into improper fractions by:

1. Multiply the denominator by the integer.

2. Add product to the numerator to get a new numerator.

3. Place the new numerator over the denominator.

Then add or subtract these improper fractions as you would any other fractions by:

1. Get a common denominator by multiplying both denominators.

2. Multiply each numerator with the other original denominator.

3. Add or subtract new numerators and put over new denominator.

Let's try this by working examples:

Example: What is 2 1/2 + 3 1/4 ?

First we can SEE the two mixed numbers using pizzas:

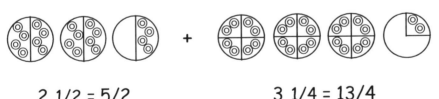

2 1/2 = 5/2   3 1/4 = 13/4

Next we will show how the math works.

Convert the mixed numbers into improper fractions:

2nd, +

$$2\frac{1}{2} = \frac{(2 \times 2) + 1}{2} = \frac{4 + 1}{2} = \frac{5}{2} \qquad \text{We see, } 2\frac{1}{2} = \frac{5}{2}$$

1st, ×

2nd, +

$$3\frac{1}{4} = \frac{(4 \times 3) + 1}{4} = \frac{12 + 1}{4} = \frac{13}{4} \qquad \text{We see, } 3\frac{1}{4} = \frac{13}{4}$$

1st, ×

Now we **add** these improper fractions, 5/2 + 13/4, by finding a common denominator, **adding** the numerators, and putting the sum over the common denominator:

$$\frac{5}{2} \diagup\!\!\!\times\!\!\!\diagdown \frac{13}{4} = \frac{(5 \times 4) + (13 \times 2)}{2 \times 4} = \frac{20 + 26}{8} = \frac{46}{8} = \frac{23}{4}$$

Therefore, 2 1/2 + 3 1/4 = 5/2 + 13/4 = 23/4.

We can write 23/4 as 5 3/4 by dividing 23 by 4, so: $\dfrac{23}{4} = 5\ 3/4$

We can also show our addition using pizzas if we slice the 1/2's into 1/4's, we see 23/4. This is 23 quarters of pizza.

Finally, see that:

2 1/2 + 3 1/4 = 5/2 + 13/4 = 10/4 + 13/4 = 23/4 = 5 3/4

Example: What is 3 1/2 - 3 1/4 ?

We can show the two mixed numbers using our pizzas:

$$3\ 1/2 = 7/2 \qquad\qquad 3\ 1/4 = 13/4$$

First convert the mixed numbers into improper fractions:

2nd, +
1st, ×

$$3\frac{1}{2} = \frac{(2 \times 3) + 1}{2} = \frac{6 + 1}{2} = \frac{7}{2} \qquad \text{We see, } 3\frac{1}{2} = \frac{7}{2}$$

2nd, +
1st, ×

$$3\frac{1}{4} = \frac{(4 \times 3) + 1}{4} = \frac{12 + 1}{4} = \frac{13}{4} \qquad \text{We see, } 3\frac{1}{4} = \frac{13}{4}$$

Now we **subtract** these improper fractions, 7/2 - 13/4, by finding a common denominator, **subtracting** the numerators, and putting the difference over the common denominator:

$$\frac{7}{2} \quad \frac{13}{4} = \frac{(7 \times 4) - (13 \times 2)}{2 \times 4} = \frac{28 - 26}{8} = \frac{2}{8} = \frac{1}{4}$$

Therefore, 3 1/2 - 3 1/4 = 7/2 - 13/4 = 1/4

We can show this using pizzas if we slice the 1/2's into 1/4's:

$$3\ 1/2 - 3\ 1/4\ =\ 7/2 - 13/4\ =\ 14/4 - 13/4\ =\ 1/4$$

Multiplying and Dividing Mixed Numbers

To multiply or divide mixed numbers:

First convert the mixed numbers into improper fractions:

1. Multiply the denominator by the integer.
2. Add product to the numerator to get a new numerator.
3. Place the new numerator over the denominator.

Then multiply or divide the improper fractions as fractions.

To **multiply fractions:**

1. Multiply the numerators.
2. Multiply the denominators.
3. Place product of numerators over product of denominators.

To **divide fractions:**

1. Flip the divisor fraction upside-down, making it a reciprocal.
2. Then multiply fractions as usual by multiplying
   the (new) numerators and the (new) denominators.

Example: If your garden is 2 1/2 meters by 3 1/4 meters, what is its size in square meters?

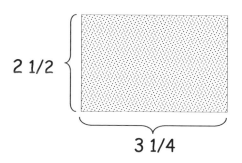

2 1/2

3 1/4

Remember that
area is:
length x width

Let's do the math...

First we convert the mixed numbers into improper fractions:

2nd, +

$$2\frac{1}{2} = \frac{(2 \times 2) + 1}{2} = \frac{4 + 1}{2} = \frac{5}{2} \qquad \text{We see, } 2\frac{1}{2} = \frac{5}{2}$$

1st, ×

2nd, +

$$3\frac{1}{4} = \frac{(4 \times 3) + 1}{4} = \frac{12 + 1}{4} = \frac{13}{4} \qquad \text{We see, } 3\frac{1}{4} = \frac{13}{4}$$

1st, ×

Our improper fractions are 5/2 and 13/4. We can just multiply the numerators and multiply the denominators:

$$\frac{5}{2} \times \frac{13}{4} = \frac{5 \times 13}{2 \times 4} = \frac{65}{8}$$

We can write 65/8 as 8 1/8 by dividing 65 by 8, so: $\frac{65}{8} = 8\ 1/8$

So, $2\frac{1}{2} \times 3\frac{1}{4} = \frac{5}{2} \times \frac{13}{4} = \frac{65}{8} = $ **8 1/8 square meters.**

Example: Divide 2 1/2 pizzas by 3 1/4 pizzas.

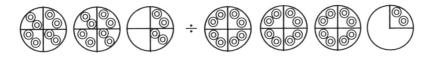

Remember, division calculates how many times one number or fraction "fits" into another number or fraction.

In this example, how many times does 3 1/4 fit into 2 1/2?

Looking at the pizzas, see 2 1/2 = 5/2 = **10/4** and 3 1/4 = **13/4**.

We see that the number of times 13/4 fits into 10/4 must be less than 1.

But first let's show the math conversion from mixed numbers, 2 1/2 and 3 1/4, into improper fractions:

$$2\frac{1}{2} = \frac{(2 \times 2) + 1}{2} = \frac{4 + 1}{2} = \frac{5}{2} \quad \text{and} \quad 3\frac{1}{4} = \frac{(4 \times 3) + 1}{4} = \frac{12 + 1}{4} = \frac{13}{4}$$

To do the math of dividing 5/2 (which equals 10/4) by 13/4, we first flip the 2nd fraction upside-down. Then multiply as usual by multiplying the (new) numerators and the (new) denominators.

$$\frac{5}{2} \div \frac{13}{4} = \frac{5}{2} \times \frac{4}{13} = \frac{5 \times 4}{2 \times 13} = \frac{20}{26} = \frac{10}{13}$$

Therefore, $2\frac{1}{2} \div 3\frac{1}{4} = \frac{5}{2} \div \frac{13}{4} = \frac{5}{2} \times \frac{4}{13} = \frac{20}{26} = \frac{10}{13}$

When we first looked at the pizzas above, we could see that 3 1/4 = **13/4** is greater than 2 1/2 = **10/4**.

This means 13/4 must fit into 10/4 less than 1 time.

Our calculated answer of 10/13 times is less than 1.

5.6. Complex Fractions

A complex fraction is a fraction that has fractions in it!
It is a fraction that has a fraction in its numerator
or in its denominator or in both.

The following are complex fractions:

$$\frac{1/5}{2}\ , \qquad \frac{5}{2/5}\ , \qquad \frac{1/6}{2/7}\ , \qquad \frac{3+1/3}{4-2/5}$$

You can simplify a complex fraction by dividing each of the
fractions. Or you can multiply the top and bottom by a
number that eliminates the embedded fraction. Let's see:

Example: Simplify (1/5)/2.

We can write $\dfrac{1/5}{2}$ as: $\dfrac{1}{5}\div 2\ =\ \dfrac{1}{5}\div\dfrac{2}{1}\ =\ \dfrac{1}{5}\times\dfrac{1}{2}\Big)\ =\ \dfrac{1}{10}$

Example: Simplify 5/(2/5).

We can write $\dfrac{5}{2/5}$ as: $\dfrac{5}{1}\div\dfrac{2}{5}\ =\ \dfrac{5}{1}\times\dfrac{5}{2}\Big)\ =\ \dfrac{25}{2}$

Example: Simplify (1/5)/2 by multiplying top and bottom by
same number to eliminate the fraction.

$$\frac{1/5}{2}\times\frac{5}{5}\ =\ \frac{\frac{1}{5}\times 5}{2\times 5}\ =\ \frac{\frac{1}{5}\times\frac{5}{1}}{2\times 5}\ =\ \frac{\frac{5}{5}}{2\times 5}\ =\ \frac{1}{10}$$

5.7. Compare Fractions: Which is Larger?

When we ask which of two fractions is larger or greater, we mean which has the larger or greater total value.

The symbol for **greater than** is > and for **greater or equal** is ≥
The symbol for **less than** is <   and for **less or equal** is ≤

1/2 is larger, or greater, than 1/4:

          1/2 > 1/4

1/2 is smaller, or less, than 3/4:

          1/2 < 3/4

Two fractions can be easily compared with each other if they have the same denominator!

The two fractions 3/4 and 2/4 can be drawn as:

          and

3/4   is larger than   2/4

Since 3 is more than 2, see that 3 fourths is more than 2 fourths.

If you need to compare two fractions that do not have the same denominator, first find a common denominator. Then you can <u>compare their numerators</u> to see which is greater.

Example: Which is larger 3/4 or 5/8?

First find the common denominator. You may notice that by multiplying both the numerator and denominator of 3/4 by 2, you have: 3/4 = 6/8. So we can compare 6/8 with 5/8 and see that 6/8 is larger than 5/8.

    >

If we find the common denominator using the same method we use when we add fractions, we can compare the numerators to see which is larger:

$$\frac{\text{one new numerator versus (vs) other new numerator}}{\text{a common denominator}}$$

$$\frac{3}{4} \; \underset{\times}{\text{VS}} \; \frac{5}{8} \longrightarrow \frac{3 \times 8 \text{ vs } 5 \times 4}{4 \times 8} \longrightarrow \frac{24 \text{ vs } 20}{32} \longrightarrow \frac{24}{32} \text{ VS } \frac{20}{32}$$

We can clearly see 24 is greater than 20, so: $\dfrac{24}{32} > \dfrac{20}{32}$

Since $\dfrac{24}{32} = \dfrac{3}{4}$ and $\dfrac{20}{32} = \dfrac{5}{8}$ then, $\dfrac{3}{4} > \dfrac{5}{8}$

If you have 3 or more fractions to compare in size, you can compare 2 fractions, then compare the 3rd with the greater of the first two. If the fractions you are comparing do not have the same denominator, first find a common denominator. Then you can <u>compare their numerators</u> to see which is bigger.

Example: Which is bigger 2/3, 3/4, or 4/5?

First compare 2/3 vs 3/4. Find the common denominator:

$$\frac{2}{3} \text{ VS } \frac{3}{4} \quad\longrightarrow\quad \frac{2\times4 \text{ vs } 3\times3}{3\times4} \quad\longrightarrow\quad \frac{8 \text{ vs } 9}{12} \quad\longrightarrow\quad \frac{8}{12} \text{ VS } \frac{9}{12}$$

We can see 9 is greater than 8, so: $\dfrac{9}{12} > \dfrac{8}{12}$

Since $\dfrac{8}{12} = \dfrac{2}{3}$ and $\dfrac{9}{12} = \dfrac{3}{4}$ then, $\dfrac{3}{4} > \dfrac{2}{3}$

We found 3/4 is larger than 2/3. We now compare 3/4 with 4/5:

$$\frac{3}{4} \text{ VS } \frac{4}{5} \quad\longrightarrow\quad \frac{3\times5 \text{ vs } 4\times4}{4\times5} \quad\longrightarrow\quad \frac{15 \text{ vs } 16}{20} \quad\longrightarrow\quad \frac{15}{20} \text{ VS } \frac{16}{20}$$

We can see 16 is greater than 15, so: $\dfrac{16}{20} > \dfrac{15}{20}$

Since $\dfrac{15}{20} = \dfrac{3}{4}$ and $\dfrac{16}{20} = \dfrac{4}{5}$ then, $\dfrac{4}{5} > \dfrac{3}{4}$

Putting $\dfrac{3}{4} > \dfrac{2}{3}$ together with $\dfrac{4}{5} > \dfrac{3}{4}$, **we see:** $\dfrac{4}{5} > \dfrac{3}{4} > \dfrac{2}{3}$

5.8. Practice Problems

5.1

(a) Use fraction reduction to show that 50 pennies have the same value as 10 nickels or 2 quarters.

(b) Can you reduce 48/12?

(c) Reduce 42/112.

5.2

(a) You live 4/5 mile from school. Your new friend says he lives 2/3 as far from school as you do. How far does he live from school?

(b) You stand 1 meter from a wall and take 10 steps toward the wall. The length of each step is 1/2 of your remaining distance from the wall. How far from the wall will you end up?

(c) Multiply : $\dfrac{1}{2} \times \dfrac{2}{3} \times \dfrac{3}{4} \times \dfrac{4}{5} \times \dfrac{5}{6} \times \dfrac{6}{7} \times \dfrac{7}{8} = ?$

5.3

(a) You want to fill an empty 1-liter container 3/4 full with water. You have a glass that holds 50 milliliters, which is 1/20 of a liter. How many full glasses should you pour into the container?

(b) You have 1/2 acre of land. You have 4 adult children, and each child has 6 children of their own. If you divide the land equally among your children, and then your children divide their portions equally among their children, how much land will each grandchild receive?

(c) Compute $\dfrac{1}{2} \div \dfrac{1}{3} \times \dfrac{1}{4} \div \dfrac{1}{5} \times \dfrac{1}{6} = ?$

5.4

(a) If you walk 2/3 mile each way to work and back each day in a 5-day work week, how many miles do you walk per week? Express your answer as an improper fraction and as a mixed number.

(b) How many 50 milliliter glasses of water are needed in order to fill 2 1/4 one-liter containers? Remember 50 ml = 1/20 liter.

(c) Show that 4 4/7 = 96/21.

5.5

(a) Sven owns 3 farms containing 4 1/2, 5 3/4, and 7 3/4 acres. Ole's farm is 18 acres. Who owns more land?

(b) Your suitcase must weigh no more than 25 lb to avoid an extra charge from the airline. The clerk puts it on a scale and it weighs 29 lb 15 oz. You remove your 1 1/4 lb camera, a 12 oz box of chocolates, a 1 3/8 lb book, and 1 1/2 lb of clothing, and stuff them into your carry-on bag. Have you removed enough weight from your suitcase to avoid the extra charge? (An ounce is 1/16 of a pound.)

(c) A giant dump truck carries a level load of dirt that is 5 3/4 yards wide, 8 1/4 yards long, and 2 1/2 yards high. How many cubic yards does it carry? (Volume = Length x Width x Height)

(d) Compute 1/2 ÷ 1/3 ÷ 4 ÷ 1/5 = ?

5.6 (a) Compute $\dfrac{1/6}{2/7}$ = ?   **(b)** Compute $\dfrac{3 + 1/3}{4 - 2/5}$ = ?

5.7

(a) Which is larger, 3 2/3 or 44/12?

(b) Which is smaller, 1/100 or 1/101?

Answers to Chapter 5 Practice Problems

5.1

(a) Remember, 50 pennies = 50 cents = a half of a dollar.
There are 100 cents in a dollar, so we can write 50 pennies as:

$$\frac{50 \text{ pennies}}{100 \text{ pennies/dollar}} = \frac{50}{100} \text{ dollars}$$

Let's look at pennies and nickels:

If we reduce $\frac{50}{100}$ by a factor of 5, (pennies per nickel):

$$\frac{50 \text{ pennies}}{100 \text{ pennies/dollar}} \frac{\div 5 \text{ pennies/nickel}}{\div 5 \text{ pennies/nickel}} = \frac{10 \text{ nickels}}{20 \text{ nickels/dollar}}$$

There are 10 nickels in 50 pennies.
Also, note there are 20 nickels in 100 pennies or in a dollar.

Let's look at pennies and quarters:

If we reduce $\frac{50}{100}$ by a factor of 25, (pennies per quarter)

$$\frac{50 \text{ pennies}}{100 \text{ pennies/dollar}} \frac{\div 25 \text{ pennies/quarter}}{\div 25 \text{ pennies/quarter}} = \frac{2 \text{ quarters}}{4 \text{ quarters/dollar}}$$

There are 2 quarters in 50 cents.
Also, note there are 4 quarters in 100 pennies or in a dollar.

(b) Yes. You can reduce 48/12 by factors of 2 and 3, or just remember that 12 is a factor of 48.
Begin dividing by 2 and keep reducing:

$$\frac{48 \div 2}{12 \div 2} = \frac{24}{6} \quad \text{Then} \quad \frac{24 \div 2}{6 \div 2} = \frac{12}{3} \quad \text{Finally} \quad \frac{12 \div 3}{3 \div 3} = \frac{4}{1} = \mathbf{4}$$

Or begin dividing by 12 which requires no further reducing:

$$\frac{48 \div 12}{12 \div 12} = \frac{4}{1} = \mathbf{4}$$

(c) Both the numerator and denominator are even, so you know you can reduce 42/112 by 2:

$$\frac{42 \div 2}{112 \div 2} = \frac{21}{56}$$

Now look for another common factor, starting with 3 and going to higher prime numbers. Neither 3 nor 5 are common, so try 7:

$$\frac{21 \div 7}{56 \div 7} = \frac{3}{8} \quad \text{which requires no further reducing.}$$

5.2

(a) This asks, "What is 2/3 of 4/5?" You multiply 4/5 by 2/3:

$$\frac{4}{5} \times \frac{2}{3} = \frac{4 \times 2}{5 \times 3} = \frac{8}{15}$$

So he lives 8/15 miles from school.

(b) The first step is 1/2 distance to the wall. After each step you are 1/2 the distance you were before, so after 10 steps you will be 1/2 multiplied 10 times.

$$\frac{1}{2} \leftrightarrow \times \leftrightarrow \frac{1}{2} \leftrightarrow \times \leftrightarrow \frac{1}{2} \leftrightarrow \times \leftrightarrow \frac{1}{2} \leftrightarrow \times \leftrightarrow \frac{1}{2} \leftrightarrow \times \leftrightarrow \frac{1}{2} \leftrightarrow \times \leftrightarrow \frac{1}{2} \leftrightarrow \times \leftrightarrow \frac{1}{2} \leftrightarrow \times \leftrightarrow \frac{1}{2} \leftrightarrow \times \leftrightarrow \frac{1}{2} =$$

$$\frac{1 \times 1 \times 1 \times 1 \times 1 \times 1 \times 1 \times 1 \times 1 \times 1}{2 \times 2 \times 2 \times 2 \times 2 \times 2 \times 2 \times 2 \times 2 \times 2} = \frac{1}{1,024} \text{ meter}$$

This is slightly less than 1 millimeter!

(c) You can set up the multiplication and cancel like numbers:

$$\frac{1 \times \cancel{2} \times \cancel{3} \times \cancel{4} \times \cancel{5} \times \cancel{6} \times \cancel{7}}{\cancel{2} \times \cancel{3} \times \cancel{4} \times \cancel{5} \times \cancel{6} \times \cancel{7} \times 8} = \frac{1}{8}$$

5.3

(a) How many 1/20s fit into 3/4?

$$\frac{3}{4} \div \frac{1}{20} = \frac{3}{4} \times \frac{20}{1} = \frac{60}{4} = \frac{30}{2} = 15 \text{ glasses fit into 3/4 liter.}$$

(b) $\frac{1}{2}$ acre $\div 4 \div 6 = \frac{1}{2} \div \frac{4}{1} \div \frac{6}{1} = \frac{1}{2} \times \frac{1}{4} \times \frac{1}{6} = \frac{1}{48}$ acre.

(c) Given $\frac{1}{2} \div \frac{1}{3} \times \frac{1}{4} \div \frac{1}{5} \times \frac{1}{6} = ?$ We flip the fractions that divide:

$$\frac{1}{2} \times \frac{3}{1} \times \frac{1}{4} \times \frac{5}{1} \times \frac{1}{6} = \frac{15}{48} \quad \text{which reduces to} \quad \frac{5}{16}$$

5.4

(a) You walk 2/3-mile twice a day for 5 days, or 10 times, so:

$$\frac{2}{3} \text{ mile} \times 10 = \frac{2}{3} \times \frac{10}{1} = \frac{20}{3} \textbf{ miles per week}, \text{ an improper fraction.}$$

Since 3 x 6 = 18, then 3 divides into 20 six times with 2 left

over, so $\frac{20}{3} = 6\frac{2}{3}$ **miles per week**, which is a mixed number.

(b) Convert 2 1/4 into an improper fraction: $\frac{(2 \times 4) + 1}{4} = \frac{9}{4}$ liters.

How many 1/20 fit into 9/4: $\frac{9}{4} \div \frac{1}{20} = \frac{9}{4} \times \frac{20}{1} = \frac{180}{4} = \textbf{45 glasses}$

(c) Write 4 4/7 as an improper fraction: $4\frac{4}{7} = \frac{(4 \times 7) + 4}{7} = \frac{32}{7}$

Reduce 96/21: $\frac{96 \div 3}{21 \div 3} = \frac{32}{7}$ Since $\frac{32}{7} = \frac{32}{7}$, then **4 4/7 = 96/21**

5.5

(a) Ole owns 18 acres. Write Swen's acreage as improper fractions with the same denominators:

$$4\ 1/2 = \frac{(4 \times 2) + 1}{2} = \frac{9}{2} = \frac{9 \times 2}{2 \times 2} = \frac{18}{4}$$

$$5\ 3/4 = \frac{(5 \times 4) + 3}{4} = \frac{23}{4} \qquad 7\ 3/4 = \frac{(7 \times 4) + 3}{4} = \frac{31}{4}$$

Add the fractions: $\frac{18 + 23 + 31}{4} = \frac{72}{4} = 18$ acres for Swen

So Swen and Ole own equal amounts of land.

(b) Convert to lbs with common denominator 16. (Note 16 oz/lb)
Suitcase: 29 lb 15 oz

$$= 29 \text{ lb} \times \frac{16}{16} + 15 \text{ oz} \times \frac{1}{16 \text{ oz/lb}} = \frac{(29 \times 16) + (15 \times 1)}{16} = \frac{479}{16} \text{ lb}$$

Now removed items:

$$1\ 1/4 \text{ lb} = \frac{(1 \times 4) + 1}{4} = \frac{5}{4} = \frac{5 \times 4}{4 \times 4} = \frac{20}{16} \text{ lb}$$

$$12 \text{ oz} = \frac{12 \text{ oz}}{16 \text{ oz/lb}} = \frac{12}{16} \text{ lb}$$

$$1\ 3/8 \text{ lb} = \frac{(1 \times 8) + 3}{8} = \frac{11}{8} = \frac{11 \times 2}{8 \times 2} = \frac{22}{16} \text{ lb}$$

$$1\ 1/2 \text{ lb} = \frac{3}{2} = \frac{3 \times 8}{2 \times 8} = \frac{24}{16} \text{ lb}$$

Subtract the 4 items from the starting weight and reduce:

$$\frac{479 - 20 - 12 - 22 - 24}{16} = \frac{401}{16} = 25\ 1/16 \text{ lb}$$

You're still 1 oz overweight!

(c) Transform the dimensions to improper fractions, then multiply to find volume.

$$5 \ 3/4 = 23/4 \qquad 8 \ 1/4 = 33/4 \qquad 2 \ 1/2 = 5/2$$

Volume = 23/4 × 33/4 × 5/2 =

$$\frac{23 \times 33 \times 5}{4 \times 4 \times 2} = \frac{3{,}795}{32} = \textbf{118 19/32 cubic yards}$$

(d) Flip divisor fractions and then multiply:

$$\frac{1}{2} \div \frac{1}{3} \div \frac{4}{1} \div \frac{1}{5} = \frac{1}{2} \times \frac{3}{1} \times \frac{1}{4} \times \frac{5}{1} = \frac{15}{8} = \textbf{1 7/8}$$

5.6

(a) Divide fraction in numerator by fraction in denominator:

$$\frac{1/6}{2/7} = \frac{1}{6} \div \frac{2}{7} = \frac{1}{6} \times \frac{7}{2} = \frac{7}{12}$$

(b) First convert the top and bottom into improper fractions, then divide:

Top: $3 + 1/3 = 3 \ 1/3 = \dfrac{10}{3}$ 　　　Note: 3 + 1/3 = 9/3 + 1/3 = 10/3

Bottom: $4 - 2/5 = \dfrac{4}{1} - \dfrac{2}{5} = \dfrac{20}{5} - \dfrac{2}{5} = \dfrac{18}{5}$

Finally, divide: $\dfrac{3 + 1/3}{4 - 2/5} = \dfrac{10}{3} \div \dfrac{18}{5} = \dfrac{10}{3} \times \dfrac{5}{18} = \dfrac{50}{54} = \dfrac{\textbf{25}}{\textbf{27}}$

5.7

(a) Transform 3 2/3 into an improper fraction: 3 2/3 = 11/3.
Next, find a common denominator for 11/3 vs 44/12:

$$\frac{11}{3} \text{ vs } \frac{44}{12} \longrightarrow \frac{11 \times 12 \text{ vs } 44 \times 3}{3 \times 12} \longrightarrow \frac{132 \text{ vs } 132}{3 \times 12} \longrightarrow \frac{132}{36} \text{ vs } \frac{132}{36}$$

Since 3 2/3 = 11/3 = 132/36 = 44/12, then 3 2/3 = 44/12. The
mixed number and fraction are the same size: **neither is larger**.
(To find the common denominator, you may have also noticed that
11/3 times 4/4 = 44/12.)

(b) First find a common denominator for 1/100 and 1/101:

$$\frac{1}{100} \text{ vs } \frac{1}{101} \longrightarrow \frac{1 \times 101 \text{ vs } 1 \times 100}{100 \times 101} \longrightarrow \frac{101 \text{ vs } 100}{10,100} \longrightarrow \frac{101}{10,100} \text{ vs } \frac{100}{10,100}$$

When the denominators are the same, the fraction with the
larger numerator is the larger fraction.

Since $\frac{1}{100} = \frac{101}{10,100}$ and $\frac{1}{101} = \frac{100}{10,100}$ then

1/100 is greater than 1/101, or **1/101 is smaller than 1/100**.

A quicker solution is to notice that since the numerators are the
same, the fraction with the smaller denominator has a greater
value, just as 1/2 is obviously larger than 1/3 or 1/10.

Chapter 6
Decimals

*And, behold, I have given the children of Levi all the **tenth** in Israel for an inheritance, for their service which they serve, even the service of the tabernacle of the congregation. Numbers 18:21*

6.1. What Are Decimals?
6.2. Add Decimals
6.3. Subtract Decimals
6.4. Multiply Decimals
6.5. Divide Decimals
6.6. Round Decimals
6.7. Compare Decimals
6.8. Practice Problems

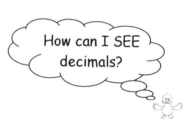

Decimals can be shown on number lines.

You can see **decimal tenths** and **decimal hundredths**:

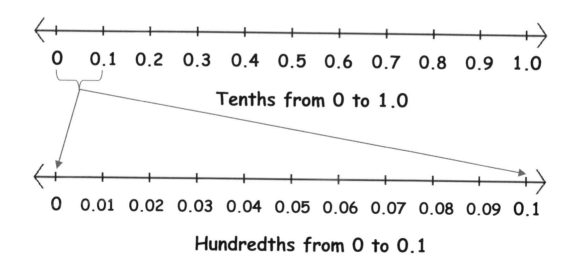

Tenths from 0 to 1.0

Hundredths from 0 to 0.1

6.1. What Are Decimals?

A decimal is another way to write a fraction.
Writing fractions as decimals can make adding, subtracting, multiplying, and dividing fractions much easier.
You can express a fraction as a decimal by dividing its numerator by its denominator:

$$2/10 = 2 \div 10 = 0.2 \qquad \text{or} \qquad 3/4 = 3 \div 4 = 0.75$$

$$10\overline{)2.0}^{\,?} \quad \rightarrow \quad 10\overline{)2.0}^{\,.2}$$

$$4\overline{)3}^{\,?} \quad \rightarrow \quad 4\overline{)3.00}^{\,.75} \\ \underline{2\,8} \\ 20$$

Decimals are fractions with denominators that are multiples of 10.

You can write 3/4 as 75/100, and 100 is a multiple of 10:

$$3/4 = 0.75 = 75/100 \quad \text{(Note } 3/4 \times 25/25 = 75/100\text{)}$$

The **decimal point** written in a decimal number is a dot or period that **separates the ones place from the tenths place.** The first digit to the right of the decimal point is the tenths place.

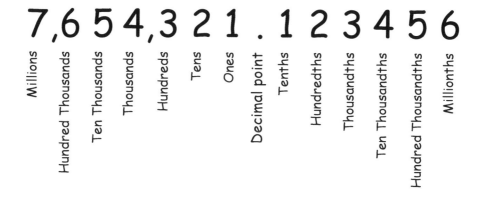

"Deci" means tenth. Decimals are **base 10** and can be written in tenths, hundredths, thousandths, and so on, to correspond with how many decimal places are in a number.

In the **base 10 system**, each digit place is 10 times greater than the digit to its right and 10 times less than the digit to its left.

Hundreds are 10 times greater than tens   (200 is 10 times 20)
Tens are 10 times greater than ones   (20 is 10 times 2)
Ones are 10 times greater than tenths   (2 is 10 times 0.2)
Tenths are 10 times greater than hundredths (0.2 is 10 times 0.02)

Digits on the right side of a decimal point are often called decimal fractions.

Examples of **decimal fractions** are:

$$0.2 , \quad 0.75 , \quad 0.0001 , \quad 0.0330$$

Decimal fractions can be written as fractions having a denominator that is a multiple of ten, such as:

$$0.75 = 75/100$$

The denominator 100 is a multiple of ten.

You can write a decimal fraction in fraction form by writing the numerator as the digits to the right of the decimal point and writing the denominator as the tenth, hundredth, or other decimal place occupied by the right-hand digit. For example:

0.2 = 2/10 = 1/5 ; **0.75** = 75/100 = 3/4 ; **0.123** = 123/1000

| tenths | hundredths | thousandths |
| place | place | place |

We can write fractions with denominators that are powers of 10, such as:

$$3/10 \quad or \quad 42/10{,}000 \quad or \quad 3{,}425/1{,}000{,}000$$

We can see decimals on the number line

The **tenths** are shown as hash marks with integers labeled:

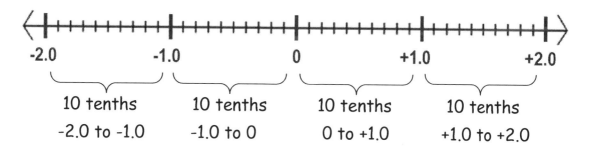

Note: 2.0 means 2 ones and 0 tenths, which equals 2.

If we label **tenths** from 0 to 1, we see (on the next number line):

0.1, 0.2, 0.3, 0.4, 0.5, 0.6, 0.7, 0.8, 0.9, 1.0

The **hundredths** are shown as hash marks with tenths labeled:

Between **0** and **+1.0** there are **100 hundredths** and **10 tenths**.

Labeling **hundredths** from **0** to **0.1** would show numbers:

0.01, 0.02, 0.03, 0.04, 0.05, 0.06, 0.07, 0.08, 0.09, 0.10

Labeling **hundredths** from **0.1** to **0.2** would show numbers:

0.11, 0.12, 0.13, 0.14, 0.15, 0.16, 0.17, 0.18, 0.19, 0.20

Let's see what the decimal tenth 0.1 looks like.

This number line shows all the tenths between 0 and 1.0 with 0 to 0.1 highlighted:

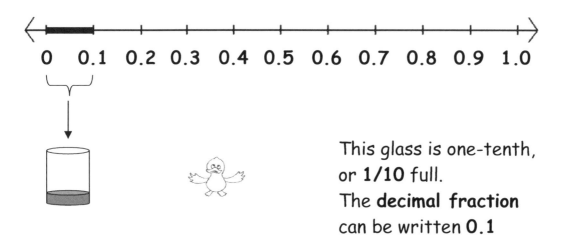

This glass is one-tenth, or **1/10** full.
The **decimal fraction** can be written **0.1**

Let's see what the decimal tenth 0.5 looks like.

This number line shows all the tenths between 0 and 1.0 with 0 to 0.5 highlighted:

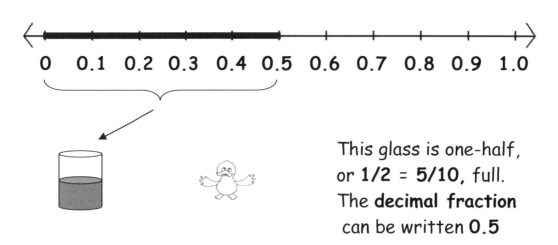

This glass is one-half, or **1/2 = 5/10**, full.
The **decimal fraction** can be written **0.5**

Decimals can be shown as decimal or regular fractions on a number line. Look at tenths, hundredths, and thousandths:

Decimal tenths from **0** to **1.0** shown as decimals and fractions:
(Tenths are the 1st place to the right of the decimal point.)

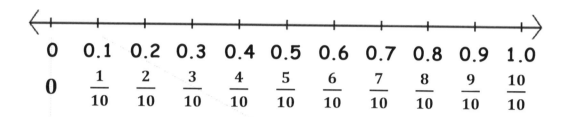

Decimal hundredths from **0** to **0.10** as decimals and fractions:
(Hundredths are the 2nd place to the right of the decimal point.)

Decimal thousandths from **0** to **0.010** as decimals and fractions:
(Thousandths are the 3rd place to the right of the decimal point.)

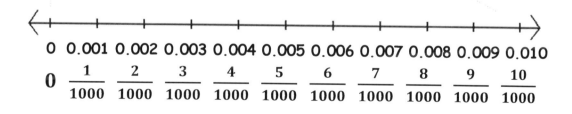

Let's identify some decimals on a number line!

Example: Write the decimals identified by the three arrows.

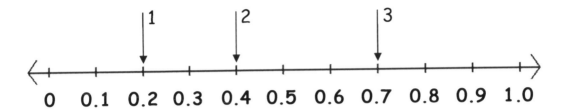

Notice that this is a number line showing **tenths** from 0 to 1.0.

Arrow 1 corresponds to 0.2 or 2/10

Arrow 2 corresponds to 0.4 or 4/10

Arrow 3 corresponds to 0.7 or 7/10

Example: Write the decimals identified by the four arrows.

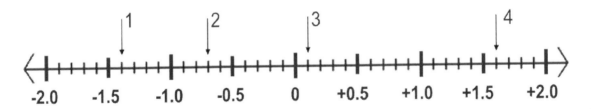

This is a number line showing tenths from -2.0 to +2.0.

Arrow 1 corresponds to -1.4 or -14/10

Arrow 2 corresponds to -0.7 or -7/10

Arrow 3 corresponds to +0.1 or 1/10

Arrow 4 corresponds to +1.6 or 16/10

Practice Writing Decimals as Numbers and Fractions

When a decimal is written as a fraction, its denominator can be some power of ten. For example, 0.2 and 0.22 can be written:

2/10   and   22/100

As a number each decimal digit is 10 times greater than the digit to its right. For example:

2,2 2 2,2 2 2 . 2 2 2 2 2 2

| Millions | Hundred Thousands | Ten Thousands | Thousands | Hundreds | Tens | Ones | Decimal point | Tenths | Hundredths | Thousandths | Ten Thousandths | Hundred Thousandths | Millionths |

Example: Write in decimal form and as a fraction: 2 tenths, 2 hundredths, 2 thousandths, 2 ten thousandths, 2 hundred thousandths, and 2 millionths.

2 tenths	= 0.2	= 2/10
2 hundredths	= 0.02	= 2/100
2 thousandths	= 0.002	= 2/1,000
2 ten thousandths	= 0.0002	= 2/10,000
2 hundred thousandths	= 0.00002	= 2/100,000
2 millionths	= 0.000002	= 2/1,000,000

Number of zeros between decimal point and 2's in decimals is one less than number of zeros in fraction denominators (underlined).

Example: Write the fractions and mixed numbers from the left-hand column below as words and as decimals.

2/10	= 2 tenths	= 0.2
3/100	= 3 hundredths	= 0.03
2 5/100	= 2 and 5 hundredths	= 2.05
101 35/100	= 101 and 35 hundredths	= 101.35
259/1,000	= 259 thousandths	= 0.259
7 17/10,000	= 7 and 17 ten thousandths	= 7.0017
8 756/100,000	= 8 and 756 hundred thousandths	= 8.00756

Note: Decimal digits are written according to their **decimal place**. For example, we can see 0.25<u>9</u> corresponds to thousandths place. 0.259 is the sum of 200 thousandths plus 50 thousandths plus 9 thousandths.

Extra Credit Example: Write 1/2 and 1/4 as decimals.

We remember that a fraction represents division.

$$1/2 = 1 \div 2 = 0.5 \qquad\qquad 1/4 = 1 \div 4 = 0.25$$

We can show the division as:

Even though you don't always see it, there is a decimal point after the ones place in any whole number.

For example, the number: 123

can be written: 123.0 or 123.00 or 123.000

Or, the number: 5,325

can be written: 5,325.0 or 5,325.0000

But you don't need the extra zeroes to the right unless you are showing that a number is really precise, since the extra zeros **suggest the number is known to that decimal place accuracy.**

If a number does not have any digits to the left of the decimal point, you just place a zero before the point.
This alerts you that there is a decimal point, since the point is small and can be difficult to see. Let's look at some examples:

Fifty-six hundredths is written: 0.56

Two tenths is written: 0.2

Nine-hundred eighty-five thousandths is written: 0.985

Let's Look at the Decimal Places in More Detail

Looking to the right of the decimal point we first get to **tenths**.
We see tenths boldface:

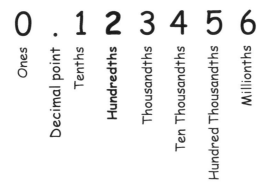

Tenths are the 1st digit after the decimal point.
We see 1 tenth = 0.1.

To the right of tenths is **hundredths**.

Hundredths are the 2nd digit after the decimal point.
We see 2 in the hundredths place. Also 0.12 is 12 hundredths.
0.12 is the sum of 10 hundredths and 2 hundredths, which equals
1 tenth and 2 hundredths since 10 hundredths = 1 tenth.

To the right of hundredths is **thousandths**.

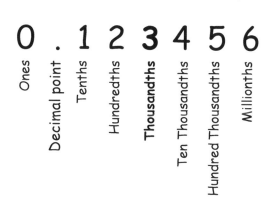

Thousandths are the 3rd digit after the decimal point.
We see 3 in the thousandths place.
Also 0.123 is 123 thousandths.
0.123 is 100 thousandths + 20 thousandths + 3 thousandths,
which equals 1 tenth and 2 hundredths and 3 thousandths.

Likewise:

Ten thousandths are the 4th digit after the decimal point.

Hundred thousandths are the 5th digit after the decimal point.

Millionths are the 6th digit after the decimal point; and so on.

Another way to visualize decimals is as a grid.

You can count the filled in regions. Look at the examples:

We see 3/10 = 3 tenths = 0.3

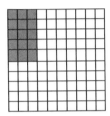

We see 15/100 = 15 hundredths = 0.15

We can show digits in a decimal number as filled boxes:

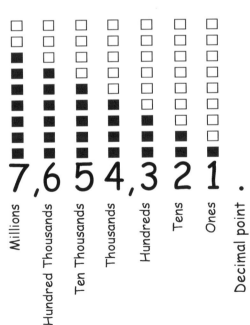

Boxes are filled in to represent the digits for 7,654,321.123456.

7,654,321 . 123456

Millions, Hundred Thousands, Ten Thousands, Thousands, Hundreds, Tens, Ones, Decimal point, Tenths, Hundredths, Thousandths, Ten Thousandths, Hundred Thousandths, Millionths

Decimals in Math, Science, Business, and Money

Decimals are used often in math, science, business, and finance.

Decimals are used to express amounts of liquid:

1 liter = 1,000 milliliters   0.001 liter = 1 milliliter
0.1 liter = 100 milliliters   0.000001 liter = 1 microliter
0.01 liter = 10 milliliters

(Prefix milli means thousandths and micro means millionth)

Decimals are used to express lengths:

0.1 meter = 1 decimeter   0.01 meter = 1 centimeter
0.001 meter = 1 millimeter   0.000001 meter = 1 micrometer
0.000000001 meter = 1 nanometer
0.0000000001 meter = 1 Angstrom (means ten-billionth meter)

(deci means tenth, centi means hundredth, nano means billionth)

Our money is expressed in decimals!

Dollars are in whole numbers, such as $1.

Cents are in tenths and hundredths, such as $0.10 and $0.01.

Let's look:

3 cents written in dollars is $0.03, which is 3/100 of a dollar

30 cents in dollars is $0.30, which is 30/100 = 3/10 of a dollar

Example: Write 201 dollars and 86 cents using decimals.

$201.86

6.2. Add Decimals

Adding decimals is just like adding whole numbers and integers, except you need to add the decimal portions too.

First let's visualize adding decimals.

Example: Add 2.5 and 3.5 using objects and number lines.

You can see **2.5 + 3.5** using diamond shapes:

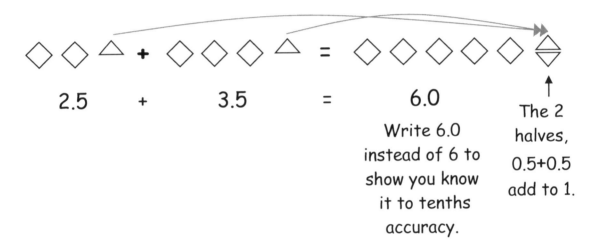

You can see **2.5 + 3.5 = 6.0** on a number line:

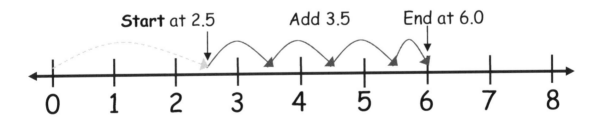

Can you think of another way to see this with number lines?

You can see 2.5 + 3.5 = 6.0 by adding two number lines:

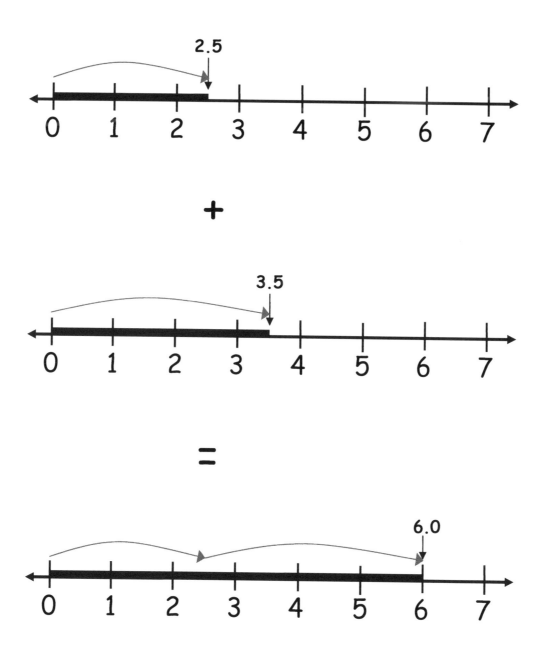

Therefore, we have seen using objects and number lines that:
2.5 + 3.5 = 6.0.

Add Decimals Using Column Format

As we just saw, adding decimals can be seen with objects and fractions of objects and also with number lines. In Section 2.1 we learned that the quickest way to add is with a column format. **Columns are a fast way to show addition of decimals.**

When you set up your columns, remember to always align the decimal points over each other so all places line up. Then begin adding from the right and move left. Carry as needed.

Example: Add 2.5 + 3.5 again, but this time use columns.

$$\begin{array}{r} 2.5 \\ + \ 3.5 \\ \hline \end{array}$$

First set up columns, aligning decimals.

$$\begin{array}{r} 1 \\ 2.5 \\ + \ 3.5 \\ \hline 0 \end{array}$$

Add tenths: 0.5 + 0.5 = 1.0.

Carry 1 of 10 tenths.

(10 tenths = 1)

$$\begin{array}{r} 1 \\ 2.5 \\ + \ 3.5 \\ \hline 6.0 \end{array}$$

Add ones column: 1 + 2 + 3 = 6.

Place decimal point in sum to right of ones place.

We have now shown **2.5 + 3.5 = 6.0** using column format.

Example: Add 5.9 + 3.08.

Set up columns aligning decimals:

$$\begin{array}{r} 5.90 \\ +\ 3.08 \\ \hline \end{array}$$

Fill in zero hundredths to the right.

$$\begin{array}{r} 5.90 \\ +\ 3.08 \\ \hline 8 \end{array}$$

Add hundredths: 0 + 8 = 8.

$$\begin{array}{r} 5.90 \\ +\ 3.08 \\ \hline 98 \end{array}$$

Add tenths: 9 + 0 = 9.

$$\begin{array}{r} 5.90 \\ +\ 3.08 \\ \hline 8.98 \end{array}$$

Add ones: 5 + 3 = 8.
Place decimal point in sum to right of ones place.

Therefore, 5.9 + 3.08 = 5.90 + 3.08 = 8.98

Now try adding 3 decimal numbers:

Example: Add 35.9 + 3.8 + 0.002.

Set up columns aligning decimals and filling in zeros to the right:

$$
\begin{array}{r}
35.900 \\
3.800 \\
+\ \ 0.002 \\
\hline
2
\end{array}
$$

After filling in a zeros, begin at right and add thousandths: 0 + 0 + 2 = 2.

$$
\begin{array}{r}
35.900 \\
3.800 \\
+\ \ 0.002 \\
\hline
02
\end{array}
$$

Add hundredths: 0 + 0 + 0 = 0.

$$
\begin{array}{r}
1\ \ \ \ \ \\
35.900 \\
3.800 \\
+\ \ 0.002 \\
\hline
702
\end{array}
$$

Add tenths: 9 + 8 + 0 = 17.

Carry the 1 in 17 tenths, which is 1.

$$
\begin{array}{r}
1\ \ \ \ \ \\
35.900 \\
3.800 \\
+\ \ 0.002 \\
\hline
9.702
\end{array}
$$

Add ones: 1 + 5 + 3 + 0 = 9.

$$
\begin{array}{r}
1\ \ \ \ \ \ \\
35.900 \\
3.800 \\
+\ \ 0.002 \\
\hline
39.702
\end{array}
$$

To add tens we just bring down the 3.

Place decimal point right of the ones place.

Therefore,

$$35.9 + 3.8 + 0.002 = 39.702$$

6.3. Subtract Decimals

Subtracting decimals is just like subtracting whole numbers and integers, except you need to subtract the decimal portion too.

First let's visualize subtracting decimals.

Example: Subtract 2.5 - 1.5 using objects and number lines.

You can see **2.5 - 1.5** using diamond shapes:

◇ ◇ △   -   ◇ △   ◇

   2.5   -   1.5   =   1.0

> Write 1.0 instead
> of 1 to show you
> know it to
> tenths accuracy.

You can see **2.5 - 1.5 = 1.0** on a number line:

End at 1.0 ← Subtract 1.5 ← **Start** at 2.5

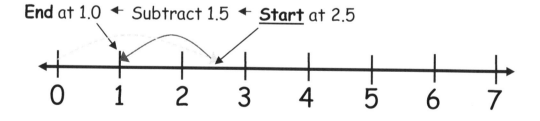

Can you think of another way to see this with number lines?

You can see 2.5 - 1.5 = 1.0 by subtracting two number lines:

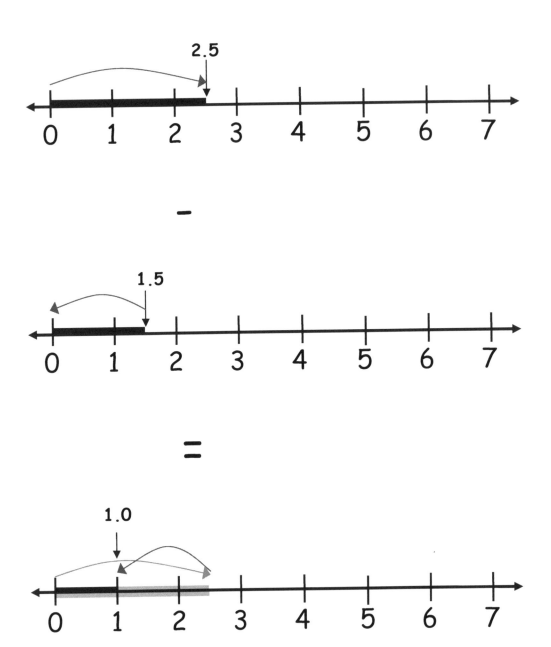

Therefore we have seen using objects and number lines that
2.5 - 1.5 = 1.0.

Subtract Decimals Using Column Format

As we just saw, subtracting decimals can be seen with objects and partial objects and also with number lines. In Chapter 2 we learned that a quick way to subtract is with column format. Likewise, **using columns is a fast way to subtract decimals.**

When you set up your columns, remember to always align the decimal points over each other so the places line up. Then begin subtracting from the right and move left. You can borrow if needed from the column to the left.

Example: Subtract 2.5 - 1.5 using columns.

$$\begin{array}{r} 2.5 \\ -\ 1.5 \\ \hline \end{array}$$

First set up columns aligning decimals.

$$\begin{array}{r} 2.5 \\ -\ 1.5 \\ \hline 0 \end{array}$$

Subtract tenths: 0.5 - 0.5 = 0.

$$\begin{array}{r} 2.5 \\ -\ 1.5 \\ \hline 1.0 \end{array}$$

Subtract ones column: 2 - 1 = 1.

Place decimal point to the right of the ones place.

We have now shown 2.5 - 1.5 = 1.0 using column format.

Now try subtracting 3 decimal numbers:

Example: Subtract 2.5 - 1.25 - 3.03.

You subtract the first two numbers. Then subtract the third from difference of the first two.

$$2.50$$
$$- 1.25$$

Fill in the zero to the right and set up columns by aligning decimals.

$$
\begin{array}{r}
4\backslash10 \\
2.\cancel{5}0 \\
- 1.2 \\
\hline
5
\end{array}
$$

We need to **borrow**.
Subtract hundredths by borrowing from the tenths resulting in: 10 - 5 = 5.

$$
\begin{array}{r}
4\backslash10 \\
2.\cancel{5}0 \\
- 1.25 \\
\hline
25
\end{array}
$$

Next subtract tenths:
4 - 2 = 2

$$
\begin{array}{r}
4\ 10 \\
2.\cancel{5}0 \\
- 1.25 \\
\hline
1.25
\end{array}
$$

Now subtract ones: 2 - 1 = 1

We see 2.5 - 1.25 = **1.25**
Next subtract the third number.

Now subtract **3.03** from **1.25**. Uh oh. 3.03 is larger than 1.25. Do you remember how to subtract a larger number from a smaller one? You first subtract the smaller number from the larger. Then take the negative. Therefore subtract 3.03 - 1.25 and then take the negative. We can check this on a **number line**.

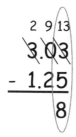

2 9/13
3.03
- 1.25
———
     8

Subtract hundredths by first borrowing from the 3 ones. Then borrow from the 10 tenths. Then subtract hundredths: 13 - 5 = 8.

2 9 13
3.03
- 1.25
———
   78

Subtract tenths: 9 - 2 = 7.

2 9 13
3.03
- 1.25
———
1.78

Subtract ones: 2 - 1 = 1.
Place decimal point to right of ones place.
Now take the negative of 1.78:
The final answer is -1.78

Therefore, 2.5 - 1.25 - 3.03 = -1.78. Look on a number line:

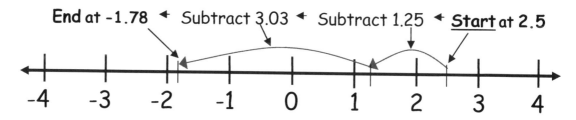

End at -1.78 ← Subtract 3.03 ← Subtract 1.25 ← **Start** at 2.5

6.4. Multiply Decimals

Remember, multiplying is a shortcut for adding the same number over and over again.

We will first look at multiplying decimals using number lines so we can visualize what is really happening.
Then we will multiply using column format.

Example: On a number line show 4.5 x 2.5.

4.5 x 2.5 is just 4.5 two and a half times! Let's show it:

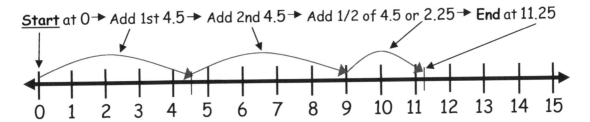

We see: 4.5 x 2.5 = 11.25.

Example: Now show 2.5 x 4.5.

2.5 x 4.5 is just 2.5 four and a half times! Let's show it:

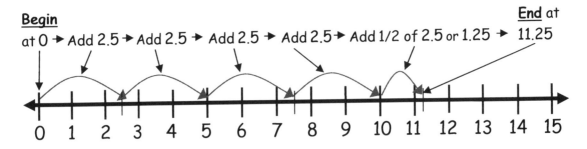

We see 2.5 x 4.5 = 11.25.   Both examples give us 11.25.

We can also show 4.5 x 2.5 using more than one number line.
To do this we can show **two and a half number lines of 4.5**:

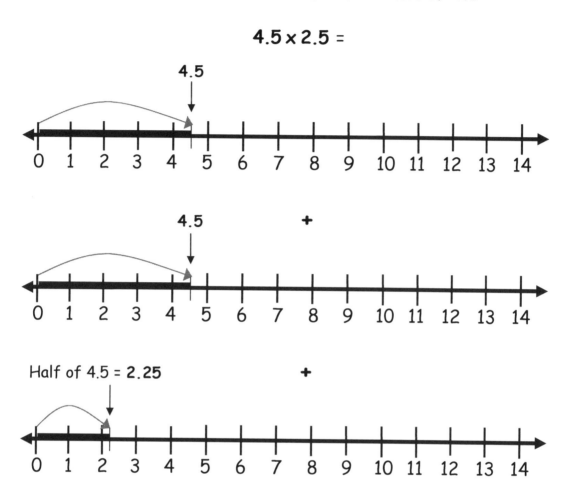

4.5 x 2.5 =

4.5

4.5 +

Half of 4.5 = **2.25** +

= 11.25

Again we see 4.5 x 2.5 = 11.25

Next we can look at this same multiplication using columns.

Multiplying decimals in columns is like multiplying whole numbers.

To multiply decimals using column format you:

1. Pretend the decimal point is not there and multiply the numbers as you would regular whole numbers. Remember to multiply each digit in the bottom number by each digit in the top number, beginning on the right and working left. Then add partial products for the multiplication answer.

2. Count the number of digits to the right of the decimal points in <u>both</u> numbers you multiplied. Place the decimal point in the product with that many digits to its right.

Example: Calculate 4.5 x 2.5 using column format.

      4.5   Set up multiplication in columns.
  x 2.5   Ignore the decimal points while multiplying.

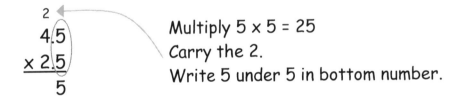

  2
  4.5   Multiply 5 x 5 = 25
x 2.5   Carry the 2.
    5   Write 5 under 5 in bottom number.

  2
  4.5   Multiply 4 x 5 = 20
x 2.5   Add the carried 2: 20 + 2 = 22
22 5   Write 22 under 2 in bottom number.
              This is the first partial product.

First partial product is **225**. Clear the carried numbers.
Next multiply by the 2 and align answer with that 2:

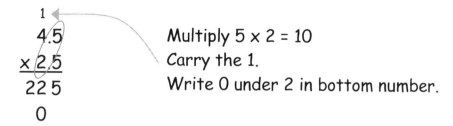

```
    1
   4.5
  x 2.5
   22 5
      0
```

Multiply 5 x 2 = 10
Carry the 1.
Write 0 under 2 in bottom number.

```
    1
   4 5
  x 2 5
   22 5
     90
```

Multiply 4 x 2 = 8
Add the carried 1: 8 + 1 = 9
Write 9 left of the 0.
This is the second partial product.

Second partial product is 900, with last (light) zero inserted for convenience. **Next add the partial products**.

```
    4.5
  x 2.5
   22 5  ⎤
   90 0  ⎦
  1125
```

Now add the partial products:
225 + 900 = 1125

Finally we **count the number of digits to the right of the decimal point in both of the numbers you multiplied**:
4.5 has 1 digit to the right of the decimal point, and
2.5 has 1 digit to the right of the decimal point.
This gives a total of 2 digits to the right of the decimal points.
Place the decimal point in the product (1125) so there are
2 digits to its right: **Therefore 4.5 x 2.5 = 11.25**.
This is the same answer we got using number lines!

Example: Multiply 0.02 by 1.23

$$
\begin{array}{r}
1.23 \\
\times\ 0.02 \\
\hline
\end{array}
$$

Set up multiplication in columns.
Choose to put 0.02 as bottom number.
Ignore the decimals while multiplying.

$$
\begin{array}{r}
1.23 \\
\times\ 0.02 \\
\hline
6
\end{array}
$$

Multiply 3 x 2 = 6
Write 6 under 2 in bottom number.

$$
\begin{array}{r}
1.23 \\
\times\ 0.02 \\
\hline
46
\end{array}
$$

Multiply 2 x 2 = 4
Write 4 to left of 6.

$$
\begin{array}{r}
1.23 \\
\times\ 0.02 \\
\hline
2\ 46
\end{array}
$$

Multiply 1 x 2 = 2
Write 2 to left of 4.
This is the first partial product.

First partial product is 246. The next number to multiply by is 0, which would give a **second partial product of 0.**
Adding partial products 246 + 0 gives 246.
Now we **count the number of digits to the right of the decimal point in both of the numbers we multiplied:**
0.02 has 2 digits to the right of the decimal point, and
1.23 has 2 digits to the right of the decimal point.
This gives a total of 4 digits to the right of the decimal points.
Place the decimal point in the product (246) so there are
4 digits to its right: Therefore, 0.02 x 1.23 = 0.0246.

6.5. Divide Decimals

Remember from Section 3.2, **division is the "opposite" of multiplication**. That means if we multiply:

2 x 3 = 6, then the opposite is 6 ÷ 2 = 3 or 6 ÷ 3 = 2

Remember division is also written as a fraction: 6/3 = 2

Division calculates how many times one number is present, or "fits", into another number.

As a reminder, look at 6 ÷ 3, or 6/3:

6 ÷ 3 asks how many 3's, or groups of 3,
fit within 6.
We see 2 groups of 3 dots fit within 6 dots.
6 ÷ 3 = 2   therefore 3 x 2 = 6

This is also true when decimals are involved.

Let's look at 7.5 ÷ 3, or 7.5 / 3:

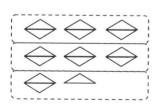

7.5 ÷ 3 asks how many 3's, or groups of 3,
fit within 7.5.
We see 2 and a half, or 2.5, groups
of 3 diamonds fit within 7.5 diamonds.
Therefore, 7.5 ÷ 3 = 2.5
This means 3 x 2.5 = 7.5

Look at Division of Decimals on a Number Line

Show on a number line: 7.5 ÷ 3 = ?

Again, 7.5 / 3 = ?, asks, "How many 3s are in 7.5?"

To find out, you can start with 7.5 and subtract 3 over and over **until you get to zero.** Let's show that on the number line:

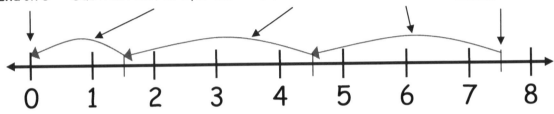

We fit two and one-half 3s before hitting zero!
Therefore, we again see that: 7.5 ÷ 3 = 2.5

We can also show 7.5 ÷ 2.5 = 3. To do this we begin at 7.5 and subtract 2.5 over and over until we get to zero.

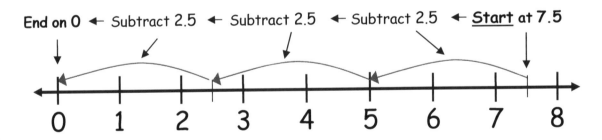

We fit three 2.5s before hitting zero!
Therefore, we see that: 7.5 ÷ 2.5 = 3

Dividing decimals with Long Division is like dividing whole numbers!

To divide decimals using long division:

1. **Put the numbers in the long division format** (Section 3.2). Remember:

Quotient ⟶ 5
Divisor ⟶ 3)16
Dividend ⟶ 15
Remainder ⟶ 1

16 ÷ 3 = 5 remainder 1
in long division
looks like this.

2. **If needed move the decimal point in the divisor to the right so there are no digits to the right of the point.** For example, 3.6 ÷ 1.2:

$$1.2)\overline{3.6}^{\ ?}$$

3. **Then move the decimal point in the dividend to the right the <u>same</u> number of places as you moved it in the divisor. Add zeros if needed.** Moving the decimal points the same number of spaces keeps the dividend and divisor the same relative size. (Like multiplying top and bottom of a fraction.) For example:

$$1.2.)\overline{3.6.}^{\ ?}$$

3.6 ÷ 1.2
becomes 36 ÷ 12

4. **Divide using the long division format procedure.**

5. **Place the decimal point in the quotient directly above where you moved it to in the dividend.**

Example: Divide 4.5 ÷ 1.5 using long division.

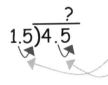

Put in long division format. Move decimal point to right in 1.5 until there are no digits to its right. This makes 1.5 become 15. Next move decimal point in 4.5 _same_ number of places, giving 45.

$$15.\overline{)45.}$$
?

Divide divisor 15 into the leftmost part of dividend it will fit into. 15 won't go into 4, but will go into 45.

$$15.\overline{)45.}$$
3.

15 goes into 45 exactly three times: 45/15 = 3. Checking gives: 3 x 15 = 45. Place the 3 above 5 in 45. Decimal point is to the right of 3.

Therefore, 4.5 ÷ 1.5 = 45 ÷ 15 = 45/15 = 3.

Example: Divide 20 ÷ 0.5 using long division.

$$0.5\overline{)20.}\;\;\;\overset{?}{}$$

Put in long division format. Move decimal point to right in 0.5 until there are no digits to its right. Moving 1 place makes 0.5 become 5. Next move decimal point in 20 the <u>same</u> 1 place.
We need to insert a zero so 20 becomes 200.

$$5.\overline{)200.}\;\;\;\overset{?}{}$$

Divide 5 into the leftmost part of 200.
5 won't go into 2, but it will go into 20.

$$5.\overline{)200.}\;\;\;\overset{4}{}$$

5 goes into 20 four times: 20/5 = 4.
Place the 4 above 20.

$$5.\overline{)200.}\;\;\;\overset{4}{}$$
$$20$$

Is there a remainder? To see first multiply the 4 by divisor 5: 4 x 5 = 20.
Put product 20 under the 20.

$$5.\overline{)200.}\;\;\;\overset{4}{}$$
$$\underline{20}$$
$$00$$

Subtract product 20 from 20 above it, resulting in the remainder of **0**: 20 - 20 = 0.

Then bring down the next part of the dividend, 0.

$$5.\overline{)200.}\;\;\;\overset{40.}{}$$
$$\underline{20}$$
$$00$$

Divisor 5 does not divide into the resulting 00, so place a 0 after 4 in the quotient. Align decimal point to the right of 0 in 40. We're done!

Therefore, 20 ÷ 0.5 = 200 ÷ 5 = 200/5 = 40.

Example: Divide 3 ÷ 8 using long division.

$$\overset{?}{8\overline{)3}}$$

Put in long division format.
Divide 8 into 3. Oops. 8 doesn't fit into 3.

$$\overset{?}{8\overline{)3.0}}$$

But 8 will go into 30. This will involve decimals!
We need to insert a zero so 3 becomes 3.0

$$\overset{3}{8\overline{)3.0}}$$

Since 8 goes into 30 three times: 30/8 = 3
plus remainder 6. Let's divide 8 into 3.0.

$$\begin{array}{r} 3 \\ 8\overline{)3.0} \\ -24 \end{array}$$

To show the remainder, first multiply 3 by
divisor 8: 3 x 8 = 24. Put 24 under the 3.0.

$$\begin{array}{r} 3 \\ 8\overline{)3.00} \\ 24 \\ \hline 6\,0 \end{array}$$

Subtract: 30 - 24 = 6, which gives remainder 6.
Now bring down the next part of the dividend,
an inserted 0.

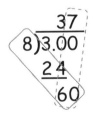

Divisor 8 divides into 60 seven times:
60/8 = 7 plus remainder 4. Let's see.

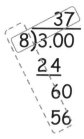

To show the remainder, first multiply 7 by divisor 8: 7 x 8 = 56. Put 56 under the 60.

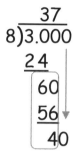

Subtract: 60 - 56 = 4, which gives remainder 4. Now bring down the next part of the dividend, an inserted 0.

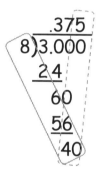

Divisor 8 divides into 40 exactly 5 times: 40/8 = 5. There is no final remainder. So we are **finished**! Remember to align decimal in answer above where it is in dividend.

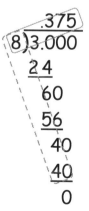

Note that if we multiplied 5 x 8 = 40 we can show: 40 - 40 = 0.

Also, in your final answer remember to put a 0 in the ones place left of the decimal point.

Therefore, 3 ÷ 8 = 0.375

6.6. Round Decimals

What is rounding?

It is slightly increasing or decreasing a number so you can approximate it with fewer digits. For example:

Rounding **52** to the nearest ten is **50**

Rounding **186** to the nearest hundred is **200**

Rounding **1 5/6** to the nearest whole number is **2**

Why Round?

1. The rounded numbers may be easier to work with.

2. To check or do quick estimates of complicated calculations.

Rounding allows you to quickly estimate an answer by first rounding the numbers you are about to add, subtract, multiply, or divide. Let's take a look at approximating an answer, then do the full calculation. For example,

$$3.9 + 2.2 = ?$$

Round to nearest whole numbers:

$$4 + 2 = ? = 6$$

Calculate **3.9 + 2.2** and compare to **estimated** answer **6**:

$$3.9 + 2.2 = 6.1$$

6.1 is very close to 6, so our estimate is pretty good!

Before we learn HOW to round decimals, let's SEE rounding:

Example: Round decimal number 2.3

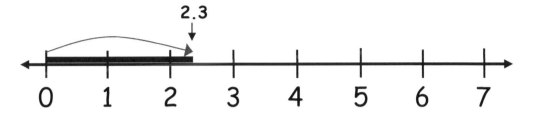

2.3 can be rounded <u>down</u> to nearest whole number, 2.

Example: Round decimal number 4.7

4.7 can be rounded <u>up</u> to nearest whole number, 5.

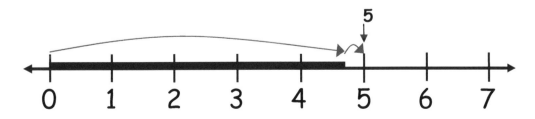

How Do You Round?

Rounding is often used to limit the number of decimal places to known accuracy. Rounding approximates the value of the decimal to the number of decimal places you choose. Here is a decimal you want to round to the tenths place. Note the digit labels.

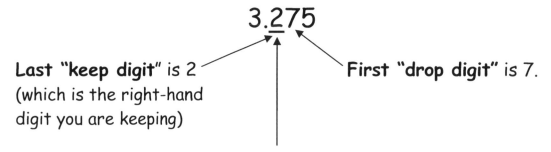

Last "keep digit" is 2
(which is the right-hand
digit you are keeping)

First "drop digit" is 7.

**When you round a number,
the last "keep digit" either gets
increased by +1 or left the same.**

Here is a Simple Rounding Method for numbers and decimals:

1. **Decide which is the last "keep digit".**

2. **If first "drop digit" is less than 5 (that is 4,3,2,1,0) "round down" by leaving last "keep digit" the same and dropping all digits to its right.** (e.g. 2.3 rounds down to 2)

3. **If first "drop digit" is 5 or greater (that is 5,6,7,8,9) "round up" by increasing the last "keep digit" by +1 and dropping all digits to its right.** (e.g. 2.6 rounds up to 3)

Example: Round 2.3 to the nearest integer using Simple Rounding.

Identify the right hand "keep digit" in 2.3 as 2.

Identify the "drop digit" in 2.3 as 3, which is less than 5.

We "round down" by leaving "keep digit" 2 the same at 2.

So, 2.3 rounds down to 2. We can see this on the number line:

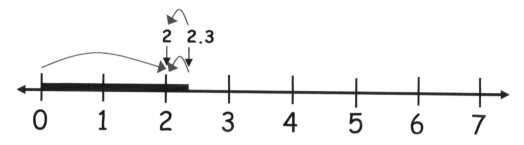

Example: Round 4.7 to the nearest integer using Simple Rounding.

Identify the right hand "keep digit" in 4.7 as 4.

Identify the "drop digit" in 4.7 as 7, which is 5 or greater.

We "round up" by increasing the "keep digit" 4 by +1 to 5.

So, 4.7 rounds up to 5. We can see this on the number line:

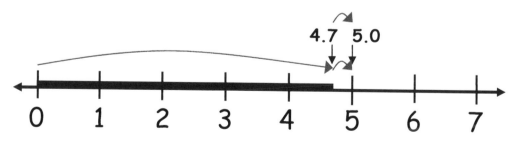

Example: Round 0.15 and 0.25 to the nearest tenth using Simple Rounding.

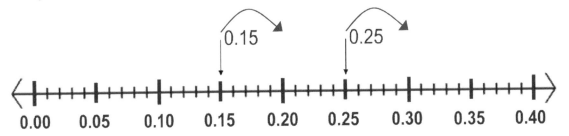

This is a number line showing **hundredths** from 0.0̲0̲ to +0.4̲0̲.

First look at 0.15:

Identify the right hand "keep digit" in 0.1̲5 as 1.

Identify the "drop digit" in 0.15̲ as 5.

Using the **Simple Rounding Method**, a first "drop digit" of 5 or greater rounds up by increasing the "keep digit" by +1.

So, 0.15 rounds up to 0.2.

Next look at 0.25:

Identify the right hand "keep digit" in 0.2̲5 as 2.

Identify the "drop digit" in 0.25̲ as 5.

Using the **Simple Rounding Method**, a first "drop digit" of 5 or greater rounds up by increasing the "keep digit" by +1.

So, 0.25 rounds up to 0.3.

Example: Round 3.275 to nearest tenth and 1,360.25468 to nearest thousandth using Simple Rounding.

Round 3.275 to tenths:

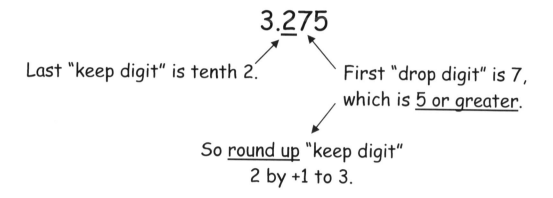

So, 3.275 rounded simply to tenth place is 3.3.

Round 1,360.25468 to thousandths:

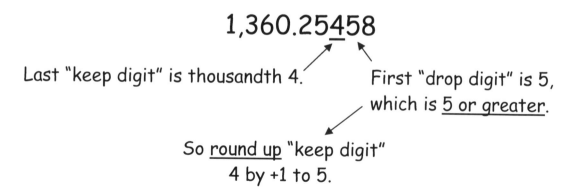

So, 1,360.25458 rounded simply to thousandth place is 1,360.255.

After a Calculation Using Physical or Scientific Measurements, You May Need to Round

If the answer you calculate is a number that has more "significant digits" than the numbers you used to calculate the answer, you need to round to the smallest number of significant digits. This is because you cannot know the answer with greater accuracy than the numbers you used to get the answer. (<u>Significant digits</u> of a number are the digits that are known with certainty.)

When you <u>multiply or divide</u>, the answer should have the same number of <u>significant digits</u> as the starting number with the fewest number of significant digits.

When you <u>add or subtract</u>, the answer should have the same number of <u>decimal places</u> as the fewest number of <u>decimal</u> places in the starting numbers.

Example: Multiply 1.23456 x 2.345 using a calculator.

If you multiply 1.23456 x 2.345 and your calculator gives 2.8950432 as the answer:

$$1.23456 \times \underline{2.345} = \underline{2.895}0432$$

You need to round to four significant digits, which is the smallest number of known significant digits in the starting numbers:

$$\underline{2.895}0432 \text{ rounded to 4 significant digits is } \underline{2.895}$$

Example: Add 1.23456 + 2.345.

Add 1.23456 + 2.345 in columns:

```
  1.23456
+ 2.34500
```
Fill in zeros.

```
  1.23456
+ 2.34500
  3.57956
```
Add columns

1.23456 + 2.345 = 3.57956

You need to round to three <u>decimal</u> places, which is the smallest number of <u>decimal</u> places in the starting numbers:

3.57956 Simply Rounded to 3 decimal places is 3.580

Example: If a container with 38.2 ounces of water has two labels, and one label says "38.2" ounces and the other label says "38" ounces, which is more accurate?

Assuming the container really has 38.2 ounces, then the label saying "38.2" describes the amount more accurately.

EXTRA CREDIT: There is a more complicated **Even/Odd Rounding Method** that is sometimes used to reduce the bias toward making numbers larger by rounding up drop digits of 5.

It is the same as the Simple Rounding Method **except when the first "drop digit" is 5**. Here's how it works:

1. If **first "drop digit" is less than 5** (that is 4,3,2,1,0) **"round down" by leaving last "keep digit" the same** and dropping all digits to its right. (Same as the Simple Method)

2. If **first "drop digit" is greater than 5** (that is 6,7,8,9) **"round up" by increasing the last "keep digit" by +1** and dropping all digits to its right. (Same as the Simple Method)

3. If **first "drop digit" is 5 and there are non-zero digits to its right, then "round up"** as you would with Simple Rounding.

4. If **first "drop digit" is 5 and it is the <u>only drop digit</u>** or all digits to its right are zero, then **if the last "keep digit" is an <u>even</u> number, "round down"** leaving last "keep digit" the same.

5. If **first "drop digit" is 5 and it is the <u>only drop digit</u>** or all digits to its right are zero, then **if the last "keep digit" is an <u>odd</u> number, "round up"** increasing the last "keep digit" by +1.

In summary, the **Even/Odd Rounding Method** is just like the Simple Rounding Method **except** when the **first "drop digit" is 5 and it is the <u>only drop digit</u>** or all **digits to its right are zero.** Then:
if the last "keep digit" is an <u>even</u> number, "round <u>down</u>", or
if the last "keep digit" is an <u>odd</u> number, "round <u>up</u>".

Example: Round 0.15 and 0.25 to the nearest tenth using Even/Odd Rounding Method.

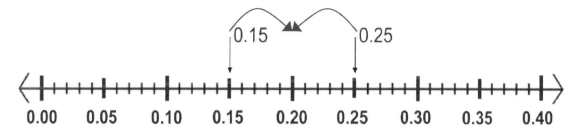

This is a number line showing **hundredths** from 0.0<u>0</u> to +0.4<u>0</u>.

First look at 0.15:

Identify the right hand "keep digit" in 0.<u>1</u>5 as 1.

Identify the "drop digit" in 0.1<u>5</u> as 5 (it is the only drop digit)

Using the Even/Odd Rounding Method,
since the last "keep digit" 1 is **odd, round up +1**.
So using Even/Odd, 0.15 rounds up to 0.2.

Note: using **Simple Rounding** a first "drop digit" of 5 or greater rounds up also, so 0.15 rounds to 0.2.

Next look at 0.25:

Identify the right hand "keep digit" in 0.<u>2</u>5 as 2.

Identify the "drop digit" in 0.2<u>5</u> as 5 (it is the only drop digit)

Using the Even/Odd Rounding Method,
since last "keep digit" 2 is **even**, we **round down.**
So using Even/Odd, 0.25 rounds down to 0.2.

Note: using **Simple Rounding** a first "drop digit" of 5 or greater rounds up, so **0.25 rounds to 0.3.**

6.7. Compare Decimals

You may need to figure out which of two or more decimals has the greater value.

To compare values you can show the decimals on a number line. For example, which decimal is greater, 0.21 or 0.12?

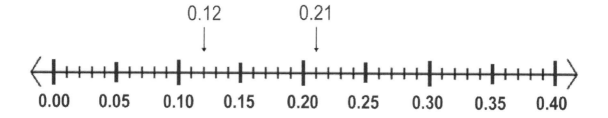

On this number line showing **hundredths** from 0.0$\underline{0}$ to +0.4$\underline{0}$, we can easily see that 0.21 is further from zero and of greater value than 0.12.

Another way to see which is greater is to write the decimals in their fraction form. Since 0.1$\underline{2}$ and 0.2$\underline{1}$ are both hundredths, we can write them as:

$$0.12 = 12/100 \quad \text{and} \quad 0.21 = 21/100$$

And ask which is larger:

$$\frac{12}{100} \quad \text{or} \quad \frac{21}{100}$$

Since 21 is larger than 12, then 21/100 is larger than 12/100.

This means: 0.21 is larger than 0.12.

Another way to see which is greater is to use columns

A quick way to see which decimal is greater is to:

1. Write decimals in a column with decimal points aligned.
2. Fill in zeros to the right of the decimal points
   so they have the same number of decimal places.
3. The greatest number will have the largest digit in the
   farthest left (greatest-value) column.

Example: Which is larger 0.000987 or 0.00123?

Write in column and align decimal points:

0.00<u>0</u>987
0.00<u>1</u>230

Largest digit in the farthest left (greatest-value) column is the 1
in the thousandths place.
So 0.00123 is larger than 0.000987.

Example: Which is larger 0.00158, 0.00089, or 0.00153?

Write in column and align decimal points:

0.00<u>15</u><u>8</u>
0.00089
0.00<u>15</u><u>3</u>

Largest digits in the farthest left (greatest-value) column are
the 1's in the thousandths places. Next-left column is two 5's.
Next-left column is the 8 and 3: Since 8 is greater than 3:
0.00158 is greater than 0.00153 is greater than 0.00089

6.8. Practice Problems

6.1 Write the following as decimals:

(a) thirty-four hundredths,

(b) thirty-four thousandths,

(c) three hundred forty millionths,

(d) thirty-four tenths, and

(e) three thousand four hundred and thirty-four hundredths.

6.2

(a) Using the method you learned in Section 4.3, add:
1/2 + 3/10 + 42/100.

(b) Using column format add 0.5 + 0.3 + 0.42.
(Note: 1/2 = 5/10 = 0.5)

(c) Use decimals to add 3 dollars, one half-dollar, three quarters, seven dimes, and 53 pennies.

(d) Add 83 ten-thousandths and 1,994 millionths.

6.3

(a) You have $10. You buy a birthday card for $3.49. How much change do you receive?

(b) Subtract 6,495 millionths from 291 ten-thousandths.

(c) You have $9.15 in your piggy bank. If you shake out 11 dimes, 14 nickels, and 107 pennies, how much is left in the pig?

6.4

(a) If a gallon of gasoline weights 6.1 pounds, and you put 12.7 gallons into your car's gas tank, how much weight do you add?

(b) Multiply 100.0 times 0.00001.

(c) EXTRA CREDIT PROBLEM: If you earn 4.25 percent simple interest per year on $168, how much interest do you earn in a year? (We'll learn percents next chapter!)

6.5

(a) Using decimals, if you evenly divide three pizzas among 8 people, what decimal fraction does each person receive?

(b) Divide 1.25 by 0.5.

(c) Now divide 0.5 by 1.25.

6.6

(a) Round 3.14159 to the nearest integer, to the nearest hundredth, and to the nearest ten-thousandth.

(b) You measure your garden at 3.79 meters by 5.8 meters. What is the most accurate number you can find for its area?

(c) You measure your garden again and have readings of 3.79 and 5.80 meters. Can you improve your calculation of its area?

6.7

(a) Which is larger in value: 10 ten-thousandths or 1,000 millionths?

(b) Which is larger in value: 589 tenths or 58,901 thousandths?

Answers to Chapter 6 Practice Problems

6.1

(a) 0.34   **(b)** 0.034   **(c)** 0.000340   **(d)** 3.4   **(e)** 34.34

6.2

(a) First:

$$\frac{1}{2} \,\, *\!\!\times\!\! \,\, \frac{3}{10} = \frac{(1 \times 10) + (3 \times 2)}{2 \times 10} = \frac{10 + 6}{20} = \frac{16}{20} = \frac{4}{5}$$

Then:

$$\frac{4}{5} \,\, *\!\!\times\!\! \,\, \frac{42}{100} = \frac{(4 \times 100)+(42 \times 5)}{5 \times 100} = \frac{610}{500} = \frac{61}{50} = 1\frac{11}{50}$$

Note: **61/50** = **1 11/50** = **1.22**

(b)

```
  0.50
  0.30
+ 0.42
  1.22
```

Our answer is **1.22**. Compared to part (a), you can see how using decimal fractions where possible can really simplify **adding** fractions!
Note: 0.22 = 22/100 = 11/50

(c)

```
  $3.00
   0.50
   0.75
   0.70
+  0.53
  $5.48
```

Write three dollars with zero cents.
One half dollar = fifty cents = $0.50.
Three quarters = seventy-five cents = $0.75.
Seven dimes = seventy cents = $0.70.
53 pennies = 53 cents = $0.53.
To get answer **$5.48** we carried the 2 in 24 tenths.

(d)

```
  0.008300
+ 0.001994
  0.010294
```

Put the 3 in the ten-thousandths place.
Put the 4 in the millionths place.
(Line up decimal points.) Answer is: **0.010294**

6.3

(a)

$$\begin{array}{r} {}^{9\ 9\ 10} \\ \$\cancel{10.00} \\ -\ 3.49 \\ \hline \$\ 6.51 \end{array}$$

Need to insert zeros and borrow.

Answer is **$6.51**

(b)

$$\begin{array}{r} {}^{8\ 10\ 9\ 10} \\ 0.029\cancel{100} \\ -\ 0.006495 \\ \hline 0.022605 \end{array}$$

Need to insert zeros and borrow.

Answer is **0.022605**

(c)

$$\begin{array}{r} \$9.15 \\ -\ 1.10 \\ \hline \$8.05 \end{array}$$

11 dimes = 11 x 10 cents = 110 cents = $1.10

Now subtract from $8.05:

$$\begin{array}{r} \$8.05 \\ -\ 0.70 \\ \hline \$7.35 \end{array}$$

14 nickels = 14 x 5 cents = 70 cents = $0.70

Now subtract from $7.35:

$$\begin{array}{r} \$7.35 \\ -\ 1.07 \\ \hline \$6.28 \end{array}$$

107 cents = $1.07 dollars

You have $6.28 left in your pig

6.4

Remember, significant digits are digits known with certainty. When you multiply or divide, the answer should have the same number of digits as the starting number with the fewest significant digits.

(a)

$$\begin{array}{r} 12.7 \\ \times\,6.1 \\ \hline 12\,7 \\ +762 \\ \hline 77.47 \end{array}$$

Move decimal point 2 places giving 77.47

Ignore the decimals and multiply: gallons x pounds/gallon = pounds. Count digits on right of decimal points in both multiplied numbers (1 + 1 = 2) and place decimal point in product with same number of digits to its right. The answer is **77.47 lbs**. Rounding to fewest significant digits, which is 2, gives **77 lbs**.

(b)

$$\begin{array}{r} 100.0 \\ \times\,0.00001 \\ \hline .001000 \end{array}$$

Move decimal point 6 places giving 0.00100

Ignoring decimals you have: 1,000 x 1 = 1,000. Count digits on right of decimal points in multiplied numbers (1 + 5 = 6) and place decimal point in product with same number of digits to its right, which gives: **0.001000**. Rounding to fewest significant digits, which is 1, gives **0.001**.

(c)

$$\begin{array}{r} 168 \\ \times\,0.0425 \\ \hline 840 \\ 336 \\ 672 \\ \hline 7.1400 \end{array}$$

Move decimal point 4 places giving 7.1400

Percent means "perCENT" or "per-hundred" or hundredths, so 4.25% = (4.25)(1/100) = 0.0425. So multiply 168 x 0.0425. Count digits on right of decimal points in multiplied numbers (0 + 4 = 4) and place decimal point in product with same number of digits to its right which gives: 7.1400. Rounded to fewest significant digits, which is 3, gives **$7.14**.

6.5

(a) We divide 8 into 3. But first recognize that 3 can be written as 3.0 or 3.000.

```
    0.375
 8)3.000
   24
    60
    56
     40
     40
      0
```

Since the divisor has no decimal part, you don't need to move the decimal in the dividend. Perform long division. Place decimal point in quotient above decimal point in the dividend. Each person gets **0.375 pizza**, which is the same as 3/8 pizza (the original division problem!).

(b)
```
     2.5
 5)12.5
   10
    25
    25
     0
```

First move the decimals one place to the right. Perform long division.
Place decimal point in the quotient directly above decimal point in the dividend, giving **2.5**.

(c)
```
        0.4
 125)50.0
     500
       0
```

First move the decimals two places to the right. Insert a zero in the dividend to complete the calculation. Perform long division.
Place decimal point in the quotient directly above decimal point in the dividend, giving **0.4**.

6.6

(a) Round to integer: 3.14159, drop digit is 1 so rounds to **3**.
Round to hundredths: 3.14159, drop digit is 1 so rounds to **3.14**.
Round to ten-thousandths: 3.14159, drop digit is 9 so rounds up
to **3.1416**. (Note: these are common estimates of Pi.)

(b) Area = length x width = 3.79 x 5.8 = 21.982. Since your least
accurate measurement 5.8 is to the nearest tenth meter and has
2 significant digits, the area (your answer) also should be
expressed with 2 significant digits. The drop number in 21.982 is
9, so 21 becomes 22. Therefore the area is **22 square meters**.

(c) 3.79 x 5.80 = 21.982. Now both initial numbers have 3
significant digits, so your answer should have 3 significant digits.
The drop digit is 8. 21.982 rounds to area **22.0 square meters**.
The answer is "more accurate" but is still 22.

6.7

(a) To compare 0.0010 and 0.001000, line up the decimal points:
0.0010
0.001000
Clearly the two numbers are equal.

(b) To compare 58.9 and 58.901, line up the decimal points:
58.9
58.901
You can see that 58.901 is larger by one thousandth.

Chapter 7
Percents

*What man of you, having an **hundred** sheep, if he lose one of them, doth not leave the ninety and nine in the wilderness, and go after that which is lost, until he find it? Luke 15:4*

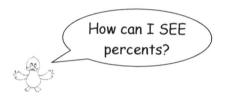

How can I SEE percents?

Think of "PER 100"

You can see and define percent as:

$$\text{PER}\underline{\text{CENT}} = \text{PER}\underline{100} = 1/\underline{100} = 0.0\underline{1}$$

↑
hundredths
place

Per<u>cent</u> is a fraction with a denominator of 100.
Percent is the number of hundredths.
Percent is another way to write decimal fractions.

7.1. What Are Percents?

Think of percent as per-hundred, or per100, or 1/100.

The symbol for percent is %.

We can show one percent, or 1%, as:

1% = 1 per<u>CENT</u> = 1 per<u>100</u> = 1/<u>100</u> = 1 hundredth = 0.0<u>1</u>

=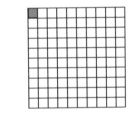

Example: Show 15% and 50% using a 100-square grid.

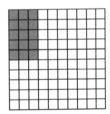

15 hundredths = 15 per100
= 15/100
= 15 perCENT = 15%

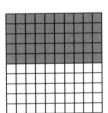

50 hundredths = 50 per100
= 50/100
= 50 perCENT = 50%

Can percent be greater than 100? Yes. For example:

100% = 100 perCENT = 100 per100 = 100/100 = 1

200% = 200 perCENT = 200 per100 = 200/100 = 2

Hint: To remember perCENT is Per100, remember a CENTury is 100 years, or 1 cent is one hundredth of a dollar.

We Can Also See perCENT Using Money

There are 100 cents (or pennies) in a dollar:

1 dollar = 100 cents and 1 cent = 1 per<u>cent</u> or 1% of a dollar

 =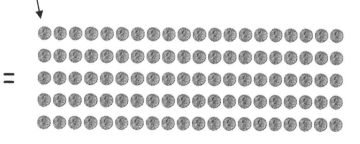

1 dollar   =   100 pennies

Example: What percent of a dollar is one quarter?

1 quarter = 25 cents = 25% or 25 per<u>cent</u> of a dollar

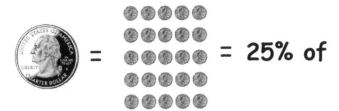 =

Example: What percent of a dollar is one dime?

1 dime = 10 cents = 10% or 10 per<u>cent</u> of a dollar

Example: If 100 people run in a race and 20 of them finish, what percent finished?

Since 20 out of 100 people finished the race,
then **20 per100** or **20 perCENT** or **20% finished**!

If 20% finish the race, then 20 out of every 100 runners finish.

> Think of per<u>cent</u> as a fraction with a denominator of 100.
> A percent can be written as a fraction or decimal.

To write a percent as a fraction:

1. Show the per<u>cent</u>, or per100, as fraction: /100

For example, **10%** is 10 per100 or $\dfrac{10}{100} = \dfrac{\cancel{10}}{\cancel{100}} = \dfrac{1}{10}$

To write a percent as a decimal:

1. First write the per<u>cent</u> as per100 or: /100

2. Then, because fraction /100 makes the number a decimal hundredth (0.01), write it as a decimal hundredth. Alternatively divide the numerator by the denominator 100, which moves the decimal point to the <u>left</u> 2 places.

For example, **10%** is 10 per100 or 10/100 = **0.1̲0̲**

Note: We can also find the decimal form by first reducing:

$\dfrac{10}{100} = \dfrac{\cancel{10}}{\cancel{100}} = \dfrac{1}{10} = 0.1\underline{0}$ ⟵ ——— 10 hundredths with the 0 in the hundredths place.

Example: Write 2% in its decimal form.

To write 2% as a decimal, first write per<u>cent</u> as per100: $\dfrac{2}{100}$

Because fraction /100, or per100, makes the number a decimal hundredth (0.01), we can write it as a decimal hundredth.

2/100 is 0.0<u>2</u> ⟵——————— See, 2 is in hundredths place

We can also divide 2 by 100 to prove 2/100 is 0.02:

```
        0.02
 100)2.00
      2 00
        0
```
Again, this shows 2% = 2/100 = 0.0<u>2</u>

Dividing by 100 moves the decimal point to the left 2 places:

So, 2% = 2 per100 = $\dfrac{2}{100}$ = .02. = **0.02**

$\div 100$

Example: Write 75% in its decimal form.

To write 75% as a decimal, first write percent as per100: $\dfrac{75}{100}$

75/100 is 75 hundredths, or: **0.75**

Let's look at the various forms in which we can write 75%:

75% = 75 per100 = $\dfrac{75}{100}$ = .75. = **0.7<u>5</u>** ⟵— hundredths place

$\div 100$

Example: Show what percent 1/4 and 1/2 represent using a 100-square grid.

For 1/4 we can use equivalent fractions to find what fraction is equivalent to 1/4 and has a denominator of 100 (for per<u>cent</u>):

$\dfrac{1}{4} = \dfrac{?}{100}$ Since 4 x 25 = 100, then $\dfrac{1 \times 25}{4 \times 25} = \dfrac{25}{100} = 25\%$

Therefore, 1/4 = 25/100

25 of the 100 boxes are filled in.

25/100 = 25 perCENT = **25%**

For 1/2 we can use equivalent fractions to find what fraction is equivalent to 1/2 and has a denominator of 100 (for per<u>cent</u>):

$\dfrac{1}{2} = \dfrac{?}{100}$ Since 2 x 50 = 100, then $\dfrac{1 \times 50}{2 \times 50} = \dfrac{50}{100} = 50\%$

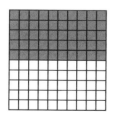

Therefore, 1/2 = 50/100

50 of the 100 boxes are filled in.

50/100 = 50 perCENT = **50%**

Example: Show what percent 2/5 and 1/50 represent using a 100-square grid.

For 2/5 we can use equivalent fractions to find what fraction is equivalent to 2/5 and has a denominator of 100 (for per<u>cent</u>):

$$\frac{2}{5} = \frac{?}{100} \qquad \text{Since } 5 \times 20 = 100, \text{ then } \quad \frac{2}{5} \frac{\times 20}{\times 20} = \frac{40}{100} = 40\%$$

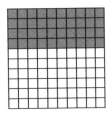

Therefore, 2/5 = 40/100

40 of the 100 boxes are filled in.

40/100 = 40 perCENT = **40%**

For 1/50 we can use equivalent fractions to find what fraction is equivalent to 1/50 and has a denominator of 100 (for per<u>cent</u>):

$$\frac{1}{50} = \frac{?}{100} \qquad \text{Since } 50 \times 2 = 100, \text{ then } \quad \frac{1}{50} \frac{\times 2}{\times 2} = \frac{2}{100} = 2\%$$

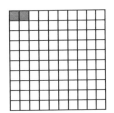

Therefore, 1/50 = 2/100

2 of the 100 boxes are filled in.

2/100 = 2 perCENT = **2%**

Example: Sam and Skippy played 5 games of chess. Sam won 3 games. What percent of the games did Skippy win?

Skippy won 5 - 3 = 2 of the 5 games, which is 2/5 of the games.
So: 2/5 is what percent, or 2/5 = ?%
Since % = per100 we write: 2/5 = ?/100. Let's see:

$$\frac{2}{5} = \frac{?}{100} \qquad \text{Since } 5 \times 20 = 100, \text{ then} \qquad \frac{2}{5} \frac{\times 20}{\times 20} = \frac{40}{100}$$

Since $\dfrac{40}{100}$ = 40 perCENT = 40%, Skippy won 40% of the games.

Example: Skippy and Sam ran 6 races. Sam won 2. What percent of the races did Skippy win?

Skippy won 6 - 2 = 4 of the 6 races, which is 4/6 of the races.
We need to find: 4/6 is what percent.
Let's find the percent by dividing 4/6:

$$
\begin{array}{r}
0.6 \\
6\overline{)4.00} \\
3\,60 \\
\hline
40
\end{array}
\quad \longrightarrow \quad
\begin{array}{r}
0.66 \\
6\overline{)4.00} \\
3\,60 \\
\hline
40 \\
36 \\
\hline
4
\end{array}
$$

This will keep going as 0.6666666666…, but we can round it to about 0.67

To get the percent, notice that **0.67 is 67 hundredths,** which is: 67 hundredths = 67 per100 = 67 perCENT = 67%
So Skippy won about 67% of the races.

7.2. Find Percents

What is Some Percent of Another Number? Let's see:

Example: What is 1% of 1 big square? Show your answer.
Remember, 1% = 1 per100 = 1/100.

$$1\% \text{ of } 1 = \frac{1}{100} \text{ of } 1 = \frac{1}{100} \times \frac{1}{1} = \frac{1\times1}{100} = 0.0\underline{1}$$

1 in hundredths place

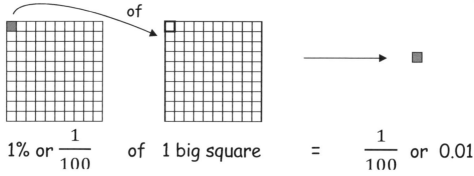

$$1\% \text{ or } \frac{1}{100} \qquad \text{of } 1 \text{ big square} \qquad = \qquad \frac{1}{100} \text{ or } 0.01$$

Example: Find and show 10% of 3 big squares. (10% = 10/100)

$$10\% \text{ of } 3 = \frac{10}{100} \text{ of } 3 = \frac{10}{100} \times \frac{3}{1} = \frac{10\times3}{100\times1} = \frac{30}{100} = 0.3\underline{0}$$

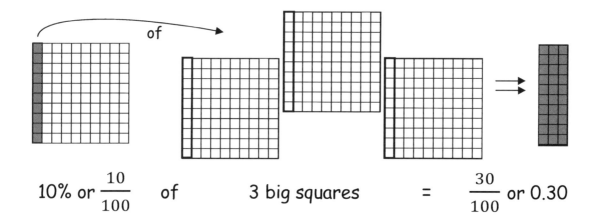

$$10\% \text{ or } \frac{10}{100} \qquad \text{of} \qquad 3 \text{ big squares} \qquad = \qquad \frac{30}{100} \text{ or } 0.30$$

We see that to find some percent of a number, we can just use the definition of a perCENT which is per100, or /100, and multiply the amount of our percent by the number.

Example: Find 6% of 15 feet? (6% = 6 per100 = 6/100.)

$$6\% \text{ of } 15 = \frac{6}{100} \text{ of } 15 = \frac{6}{100} \times \frac{15}{1} = \frac{6\times15}{100\times1} = \frac{90}{100} = 0.9\underline{0} \text{ feet}$$

$$\begin{array}{c} 15 \\ \underline{\times 6} \\ 90 \end{array}$$

90 hundredths

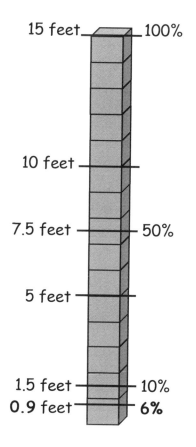

15 feet ___ 100%

10 feet

7.5 feet — 50%

5 feet

1.5 feet — 10%
0.9 feet — 6%

Here we have a 15 foot tower.
You can see that going 100% of
the way up our tower,
we have gone 15 feet.
Going 50% up our tower,
we have gone 7.5 feet.
Going 10% up our tower,
we have gone 1.5 feet.
Going half of 10%, or 5%,
up the tower would be
half of 1.5 feet, or 0.75 feet.
Since 6% is a bit more than 5%,
it makes since that
going 6% up
corresponds to 0.9 feet.

Example: Find 35% of 100? (Remember, 35% = 35/100.)

$$35\% \text{ of } 100 = \frac{35}{100} \text{ of } 100 = \frac{35}{100} \times \frac{100}{1} = \frac{35 \times 100}{100 \times 1} = \frac{3500}{100} = \mathbf{35.\underline{00}}$$

3,500
hundredths

Note: We can also cancel $\dfrac{35 \times \cancel{100}}{\cancel{100}}$ to get 35.

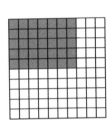

35% = 35/100
Here, 100 boxes is 100% of the boxes.
35 of the 100 boxes are filled in.

Example: Find 5% of 4 meters? (Remember, 5% = 5/100.)

$$5\% \text{ of } 4 = \frac{5}{100} \text{ of } 4 = \frac{5}{100} \times \frac{4}{1} = \frac{5 \times 4}{100 \times 1} = \frac{20}{100} = \mathbf{0.\underline{20}} \text{ meters}$$

20
hundredths

Note: We can also cancel $\dfrac{2\cancel{0}}{10\cancel{0}} = \dfrac{2}{10}$ to get 0.2

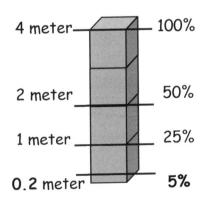

SEE that going 100% of the way up
our tower,
we have gone 4 meters.
Going 50% up, we've gone 2 meters.
Going 25% up, we've gone 1 meter.
You can see that going 5% up,
corresponds to **0.2 meters**.

Example: What are the following percents of 7: 1% of 7, 10% of 7, 100% of 7, 1,000% of 7, and 10,000% of 7.

We can use the definition of a perCENT which is /100, and multiply the amount of our percent by the number.

$$1\% \text{ of } 7 = \frac{1}{100} \text{ of } 7 = \frac{1}{100} \times \frac{7}{1} = \frac{1\times7}{100\times1} = \frac{7}{100} = \mathbf{0.0\underline{7}}$$

$$\text{7 hundredths}$$

$$10\% \text{ of } 7 = \frac{10}{100} \text{ of } 7 = \frac{10}{100} \times \frac{7}{1} = \frac{10\times7}{100\times1} = \frac{70}{100} = \mathbf{0.7\underline{0}}$$

$$\text{70 hundredths}$$
$$\text{or just } 0.7$$

$$100\% \text{ of } 7 = \frac{100}{100} \text{ of } 7 = \frac{100}{100} \times \frac{7}{1} = \frac{100\times7}{100\times1} = \frac{700}{100} = \mathbf{7.0\underline{0}}$$

$$\text{700 hundredths}$$
$$\text{or just } 7$$

$$1,000\% \text{ of } 7 = \frac{1,000}{100} \times \frac{7}{1} = \frac{1,000\times7}{100\times1} = \frac{7,000}{100} = \mathbf{70.0\underline{0}}$$

$$\text{7,000 hundredths}$$
$$\text{or just } 70$$

$$10,000\% \text{ of } 7 = \frac{10,000}{100} \times \frac{7}{1} = \frac{10,000\times7}{100\times1} = \frac{70,000}{100} = \mathbf{700.0\underline{0}}$$

$$\text{70,000 hundredths}$$
$$\text{or just } 700$$

Example: What is 350% of 1/5 of a dollar?

$$350\% \text{ of } 1/5 = \frac{350}{100} \times \frac{1}{5} = \frac{350 \times 1}{100 \times 5} = \frac{350}{500} = \frac{350 \div 5}{500 \div 5} = \frac{70}{100} = 0.70$$

70 hundredths
or just 7 tenths of a dollar
which is **70 cents**

Even though we are talking about money, we can
build a tower of blocks to SEE why 350% of 1/5
correlates with 7/10 = 7 tenths = 70 hundredths = 70 cents

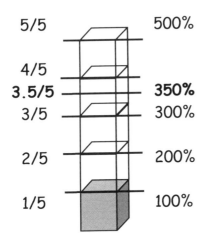

See 100% of 1/5 of the tower.
Each block is 100% of 1/5.
Going up 3 blocks is at 300%
of 1/5.
Going up 3 and 1/2, or **3.5**,
blocks is at **350%** of 1/5.

$$\frac{3.5}{5} = \frac{3.5 \times 2}{5 \times 2} = \frac{7}{10} = 0.7$$

which again shows 7 tenths of
a dollar or **70 cents**.

An alternative way to find some percent of a number is to convert the percent to its decimal form. We can convert the percent to its decimal form by either using the definition of a perCENT (/100) or by dividing. **Then we multiply that decimal by the number.** Let's see:

Example: Find 5% of 4 meters and 25% of 4 meters.

First convert the percents, 5% and 25%, to their decimal forms, then multiply:

$$5\% \text{ of } 4 = \frac{5}{100} \text{ of } 4 = 0.0\underline{5} \times 4 = 0.2\underline{0} \text{ or just } \mathbf{0.2}$$

$$25\% \text{ of } 4 = \frac{25}{100} \text{ of } 4 = 0.2\underline{5} \times 4 = 1.0\underline{0} \text{ or just } \mathbf{1}$$

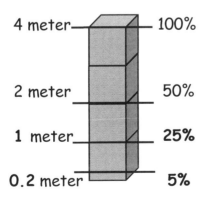

SEE that 100% of the way up our tower is 4 meters.
50% up is at 2 meters.
25% up is at 1 meter.
5% up is at 0.2 meters.

Can we also find a fraction of a percent? Yes, because it is just the reverse order of multiplication. And we learned that the order we multiply doesn't matter.

Example: Find 3/4 of 80%.

$$3/4 \text{ of } 80\% = \frac{3}{4} \times \frac{80}{100} = \frac{3 \times 80}{4 \times 100} = \frac{240}{400} = \frac{240 \div 4}{400 \div 4} = \frac{60}{100} = 60\%$$

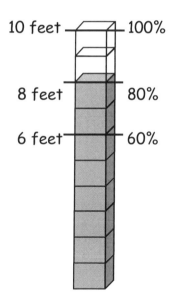

10 feet ___ 100%

8 feet ___ 80%

6 feet ___ 60%

To illustrate we have
a 10-foot tower.
100% of the way up is 10 feet.
80% of the way up is 8 feet.

$\frac{3}{4}$ of 8 feet is 6 feet,

so $\frac{3}{4}$ of 80% of the whole tower

is 60% of the whole tower.

7.3. Add and Subtract Percents

To add or subtract percents, just add or subtract the percents written in their percent forms. Let's see:

$$\% \; + \; \% \; = \; \% \qquad or \qquad \% \; - \; \% \; = \; \%$$

We can just add or subtract the percents directly because percents are just perCENT or /100 or hundredths! They are really just **fractions with a common denominator of 100** and are therefore in the **same units of hundredths.**

Example: Calculate 22% + 6% = ?

22% and 6% are both written in their percent form.
Since 22 + 6 = 28,
Then, **22% + 6% = 28%**

Example: Calculate 169% - 23% = ?

169% and 23% are both written in their percent form.
Since,

```
 169
- 23
 146
```

Then, **169% - 23% = 146%**

Example: If you have a bag of 100 mixed nuts and 10% are cashews, 23% are pecans, 5% are pistachios, 2% are walnuts, and the rest are peanuts, what percent are peanuts? How many peanuts are in the bag?

Total Nuts = 10% Cashews + 23% Pecans + 5% Pistachios + 2% Walnuts + ?% Peanuts

First find the non-peanut nuts:
10% cashews + 23% pecans + 5% pistachios + 2% walnuts =
10% + 23% + 5% + 2% = **40% are non-peanuts**

Total nuts are 100%. So total nuts are peanuts + non-peanuts.
This means 100% of nuts must be:
100% total nuts = ?% peanuts + 40% non-peanuts

To find % peanuts subtract 40% non-peanuts from 100% nuts:
100% total nuts - 40% non-peanuts = **60% peanuts in the bag**

We now need to find how many peanuts are in the bag.
We are told there are 100 total nuts.
So how many nuts are represented by 60%?
Remember the definition of percent: 60% = 60 per100,
where 60 per100 means 60 per every 100.

$$60\% \times 100 \text{ nuts} = \frac{60}{100} \times \frac{100}{1} = \frac{6{,}000}{100} = 60$$

Or we can multiply: 60% × 100 nuts = 0.60 × 100 = 60

So in a bag of 100 nuts, 60% are peanuts
So 60 per100 which is 60 nuts are peanuts!

Example: There are 50 students in your class. 20% want to study math, 25% want to study history, 35% want to study science, and the rest don't know what they want to study. What % know what they want to study and what % don't know? How many students don't know?

100% total students = % who know + % who don't know

Students who know what to study are:

20% math + 25% history + 35% science = % who know

20% + 25% + 35% = **80% who know what to study**

To find **% who don't know** what to study,
subtract 80% who know from 100% total students:

100% total students - 80% who know = **20% who don't know**

Now find **how many** students don't know what to study.
We are told there are 50 total students.
So how many students are represented by 20% of the 50?
To find this think of the definition of percent:
20% = 20 perCENT = 20 per100
There are 20 "don't knows" per 100 students.
But there are only 50 students. We can notice that
50 is 1/2 of 100 so there must be 1/2 of 20 or **10 "don't knows"**
Or we can find "don't knows" using equivalent fractions:

$$\frac{20}{100} \frac{\div 2}{\div 2} = \frac{10}{50}$$ **Therefore, there are 10 who don't know**

← 10 of 50

7.4. Multiply and Divide Percents

To multiply and divide percents,
1. Write each per<u>cent</u> as its decimal hundredth (0.01).
   Or divide each number by 100 to get its decimal form.
2. Multiply or divide the decimal as described in Chapter 6.
3. You can then convert the answer back into a percent by
   multiplying by 100 (which moves the decimal point 2 places
   to the right).

Example: Multiply 32% x 6%.

32% x 6% = 32 per100 x 6 per100 = 32/100 x 6/100 = 0.3<u>2</u> x 0.0<u>6</u>

$$\begin{array}{r} 0.32 \\ \times\ 0.06 \\ \hline 0.0192 \end{array}$$

To multiply decimals, multiply the normal
way. Then count the digits to the right of
the decimal points in both multipliers.
Place the decimal point in the product with
that many digits to its right.

hundredths

Move point
4 places

0.0<u>1</u>92 = 1.92 hundredths = 1.92/100 = 1.92% or about 2%
We can also find % from a decimal by multiplying by 100:
0.0192 x 100 = 1.92% or about 2%
Now let's look at 32% x 6% = 2%:

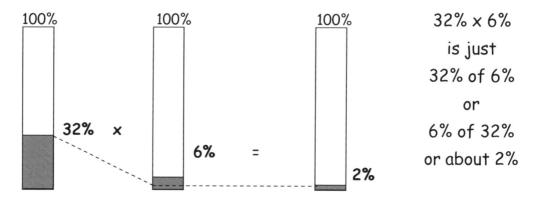

100%   100%   100%   32% x 6%
                                           is just
                                          32% of 6%
                                             or
              32%   x   6% of 32%
                     6%   =   or about 2%
                                      2%

Example: What is 20% of 70%.

$$20\% \text{ of } 70\% = 20 \text{ per}100 \times 70 \text{ per}100 = \frac{20}{100} \times \frac{70}{100} = 0.2\underline{0} \times 0.7\underline{0}$$

And: $0.2 \times 0.7 = 0.14$ ← Note: Total of 2 digits to right of the decimal points in 0.2 and 0.7 give 2 digits to right of point in 0.14

Decimal 0.1$\underline{4}$ = 14 hundredths = 14/100 = 14%

Therefore, 20% of 70% is 14%

Let's look at this:

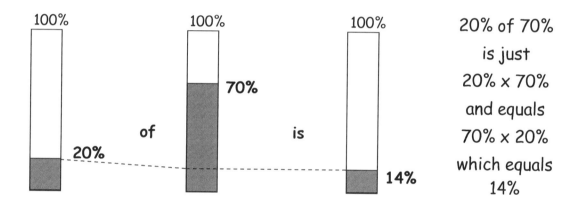

20% of 70% is just 20% x 70% and equals 70% x 20% which equals 14%

Equivalently (multiplying is commutative so order doesn't matter):

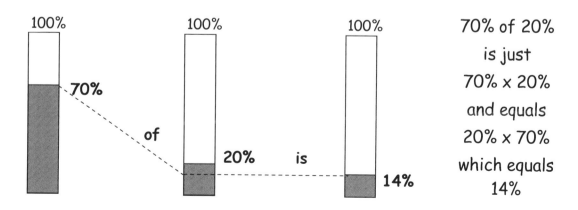

70% of 20% is just 70% x 20% and equals 20% x 70% which equals 14%

Example: Divide 20% ÷ 5%.

$$20\% \div 5\% \;=\; \frac{20}{100} \div \frac{5}{100} \;=\; 0.2\underline{0} \div 0.0\underline{5}$$

We can divide this two ways:

1. Do the division 20/100 ÷ 5/100 in fraction form:

$$\frac{20}{100} \div \frac{5}{100} \;=\; \frac{20/100}{5/100} \;=\; \frac{20/\cancel{100}}{5/\cancel{100}} \;=\; \frac{20}{5}$$

Cancel 100's

5 goes into 20
exactly 4 times
(since 5 x 4 = 20)

2. Or divide the decimals as in Section 6.5: 0.20 ÷ 0.05

$$0.05\overline{)0.20}^{\;?} \qquad \longrightarrow \qquad 5.\overline{)20.}^{\;4.}$$

Move decimal points

5 goes into 20
exactly 4 times

Therefore, 20% ÷ 5% = 20/100 ÷ 5/100 = 0.20 ÷ 0.05 = 4
Let's look at this:

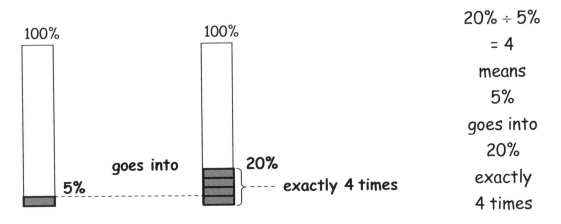

100%   100%

5%   goes into   20% exactly 4 times

20% ÷ 5%
= 4
means
5%
goes into
20%
exactly
4 times

Example: Divide 75% by 25%.

$$75\% \div 25\% = \frac{75}{100} \div \frac{25}{100} = 0.75 \div 0.25$$

We can do this two ways:

1. Do the division 75/100 ÷ 25/100 in fraction form:

$$\frac{75}{100} \div \frac{25}{100} = \frac{75/100}{25/100} = \frac{75/\cancel{100}}{25/\cancel{100}} = \frac{75}{25}$$

25 goes into 75 exactly 3 times
(since 25 x 3 = 75)

2. Or divide the decimals as in Section 6.5: 0.75 ÷ 0.25

$$0.25\overline{)0.75}^{\,?}$$ ⟶ $$25.\overline{)75.}^{\,3}$$

Move points

25 goes into 75 exactly 3 times

Therefore, 75% ÷ 25% = 75/100 ÷ 25/100 = 0.75 ÷ 0.25 = 3

Let's look:

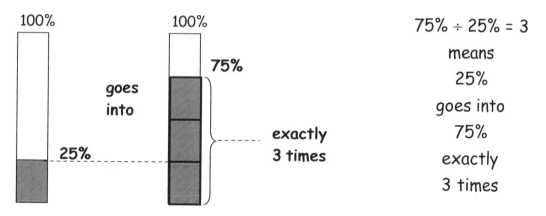

75% ÷ 25% = 3
means
25%
goes into
75%
exactly
3 times

We can also think of 0.75 ÷ 0.25 as how many quarters fit into 75 cents? Note: 0.25 = 25 cents and 0.75 = 75 cents. Yes, it's 3!

7.5. Practice Problems

7.1

(a) If you earn $100,000 and give 10% of your earnings to charity, how much will you have left?

(b) What is the difference between pennies (cents) and percent of a dollar?

7.2

(a) 65% of your class of 40 students votes for you for student council. How many votes did you receive?

(b) In an election in your town, a candidate received 110% of the votes of the 2,000 registered voters. How many votes did he receive?

7.3

At school you are learning about the financial markets.
(a) A stock price is $100. Then it goes up in value by 150%. Using addition of percentages, find its new value.

(b) A stock had been worth $400, but then lost 30% of its value. Using subtraction of percentages, find its new value.

7.4

(a) In your school you win 75% of the vote. Of those voting for you, 20% are close friends. Of all votes cast, what percent were cast by close friends for you? (Use multiplication of percents.)

(b) EXTRA CREDIT PROBLEM: Snooty answered 90% of the math questions correctly on a test. You answered only 60% correctly. Snooty brags that he answered correctly 30% more often than you did. Is Snooty correct in his bold assertion?

Answers to Chapter 7 Practice Problems

7.1

(a) First find 10% of $100,000. Remember 10% = 10 per100.

$$\frac{10}{100} \times \frac{100,000}{1} = \frac{1,000,000}{100} = \frac{10,000}{1} = \$10,000 \text{ you give away}$$

The amount you have left is: $100,000 - $10,000 = **$90,000**.

(b) They are the same. For example: One dollar = 100 pennies = 100 cents = 100 hundredths of a dollar = 100% of a dollar.
Also: 1 penny = 1 cent = 1/100 of a dollar = 1% of a dollar.
Also: 25 cents = 25% of a dollar.

7.2

(a) 65% of 40 = $\dfrac{65}{100} \times \dfrac{40}{1} = \dfrac{2,600}{100}$ = **26 votes**

(b) 110% of 2,000 = $\dfrac{110}{100} \times \dfrac{2,000}{1} = \dfrac{220,000}{100}$ = 2,200 votes

This town needs new election judges!

7.3

(a) The stock starts at $100, which is 100% of its value.
It increases by 150%. That is 100% initial + 150% increase.
Its new value is 100% + 150% = 250% of the $100 initial value.

$$250\% \text{ of } 100 = \frac{250}{100} \times \frac{100}{1} = \frac{250 \times \cancel{100}}{\cancel{100} \times 1} = \frac{250}{1} = \mathbf{\$250}$$

(b) (100% initial percent) - (30% decrease percent) = 70%
Its new value is 100% - 30% = 70% of the $400 initial value:

$$70\% \text{ of } \$400 = \frac{70}{100} \times \frac{400}{1} = \frac{28,000}{100} = \$280.$$

7.4

(a) This is asking what is 20% of your 75% of the votes:

$$20\% \text{ of } 75\% = \frac{20}{100} \times \frac{75}{100} = \frac{1,500}{10,000} = \frac{15}{100} = 15 \text{ per}100 = \textbf{15\%}$$

So of all votes cast, 15% were cast by close friends for you!

(b) No! Snooty may have gotten a score that is higher by 30% (90% - 60% = 30%), but that does not tell you what his score is **as a percent** of your score. To find this you need to know:
     what is Snooty's additional 30% as a percent of your 60%?
So to find what percent his additional 30% is of your 60%,
you divide his additional 30% by your 60%.
This tells you how many times 60% fits into 30%.

$$30\% \div 60\% = \frac{30/100}{60/100} = \frac{30/\cancel{100}}{60/\cancel{100}} = \frac{30}{60} = 0.50$$

Then you express decimal 0.50 as a percent. To do this either
see: 0.50 = 50 hundredths = 50 percent = **50%**
Or remember % = 1/100. So you multiply 0.50 by 100 since by
writing % you are effectively dividing by 100. So: 0.50 = 50%
The answer is 50%, NOT 30%! Snooty isn't so smart after all.

Chapter 8
Converting Fractions-Decimals-Percents

*The law of the Lord is perfect, **converting** the soul:*
the testimony of the Lord is sure, making wise the simple. Psalm 19:7

8.1. Convert Fractions to Decimals
8.2. Convert Fractions to Percents
8.3. Convert Decimals to Fractions
8.4. Convert Decimals to Percents
8.5. Convert Percents to Fractions
8.6. Convert Percents to Decimals
8.7. Practice Problems

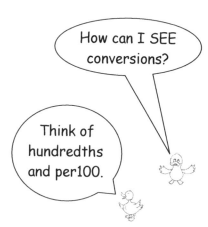

How can I SEE conversions?

Think of hundredths and per100.

Fractions→Decimals   1/100 = 0.01 ←— **hundredths place**

Fractions→Percents   1/100 = 1 PER 100 = 1 PERCENT

Decimals→Fractions   0.01 = 1/100

Decimals→Percents   0.01 = 1 PER100 = 1 PERCENT

Percents→Fractions   1 PERCENT = 1 PER100 = 1/100

Percents→Decimals   1 PERCENT = 1 PER100 = 0.01

In summary: 1% = 1 PERCENT = 1 PER100 = 1/100 = 0.01

8.1. Convert Fractions to Decimals

Fractions→Decimals

Examples:

$$\frac{1}{100} = 0.0\underline{1}$$
↑
hundredths place

$$\frac{5}{10} = 0.\underline{5}$$
↑
tenths place

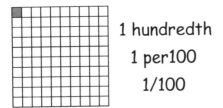

1 hundredth

1 per100

1/100

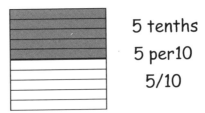

5 tenths

5 per10

5/10

A fraction is another way to write a decimal.
A decimal is another way to write a fraction.

Decimals are fractions with denominators that can be
expressed as multiples of 10, such as 10, 100, or 1,000.

For example, you can write 3/4 with a denominator of 100:

$$\frac{3}{4} = \frac{75}{100} = 0.7\underline{5}$$
↑
hundredths place

$$(\text{note} \quad \frac{3}{4} \frac{\times 25}{\times 25} = \frac{75}{100})$$

Here are two ways to write a fraction as a decimal:

1. Create an equivalent fraction with a multiple of 10 for its denominator: /10, /100, /1,000, /10,000, etc. Then because the fraction makes a number of tenths, hundredths, thousandths, etc., it may be easily written as a decimal such as 0.1, 0.01, 0.001, etc.

Or:

2. Simply divide the numerator by the denominator.

Example: Write 5/25 in its decimal form.

You can make an equivalent fraction with a denominator of 10, 100, etc. Let's choose 100 (since 25 x 4 = 100). Then write as decimal hundredth:

$$\frac{5}{25} \frac{\times 4}{\times 4} = \frac{20}{100} = 0.2\underline{0} \qquad \textbf{Therefore, 5/25 = 0.20}$$

20 in hundredths place.

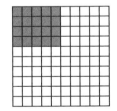

5/25 = 20/100 = 0.20

20 of the 100 boxes are filled in.

Alternatively we can divide 5 by 25 to find the decimal form:

```
      0.2
  25)5.0
      5 0
      ───
        0
```

Again, 5/25 = 0.2

Example: Write fraction 3/4 as a decimal.

You can make an equivalent fraction with a denominator of 10, 100, etc. Let's choose 100 (since 4 x 25 = 100). Then write as decimal hundredth:

$$\frac{3}{4} = \frac{3 \times 25}{4 \times 25} = \frac{75}{100} = 0.7\underline{5} \qquad \textbf{Therefore, 3/4 = 0.75}$$

↑
hundredths place

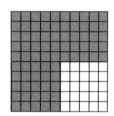

3/4 = 75/100 = 0.75

75 of the 100 boxes are filled in.

Alternatively we can divide 3 by 4 to find the decimal form:

$$
\begin{array}{r}
0.7 \\
4\overline{)3.00} \\
2\,8 \\
\hline
20
\end{array}
\qquad \rightarrow \qquad
\begin{array}{r}
0.75 \\
4\overline{)3.00} \\
2\,8 \\
\hline
20 \\
20 \\
\hline
0
\end{array}
$$

Again, we show: 3/4 = 0.75

Example: Write fraction 3/5 as a decimal.

You can make an equivalent fraction with a denominator of 10, 100, etc. Let's choose 10 (since 5 x 2 = 10). Then write as decimal tenth:

$$\frac{3}{5} = \frac{3 \times 2}{5 \times 2} = \frac{6}{10} = 0.\underset{\uparrow}{6} \qquad \textbf{Therefore, 3/5 = 0.6}$$

tenths place

3/5 = 6/10 = 0.6

6 of the 10 boxes are filled in.

Alternatively we can divide 3 by 5 to find the decimal form:

```
     0.6
  5)3.0
     3 0
       0
```

Again, we show: 3/5 = 0.6

Example: Write fraction 3/8 as a decimal.

You can make an equivalent fraction with a denominator of 10, 100, 1,000, etc. Let's choose 1,000 (since 8 x 125 = 1,000). Then write as decimal thousandth:

$$\frac{3}{8} = \frac{3 \times 125}{8 \times 125} = \frac{375}{1,000} = 0.375 \qquad \textbf{Therefore, 3/8 = 0.375}$$

↑ thousandths place

Alternatively we can divide 3 by 8 to find the decimal form:

$$\begin{array}{r} 0.3 \\ 8\overline{)3.00} \\ 2\,4 \\ \hline 60 \end{array} \quad\rightarrow\quad \begin{array}{r} 0.37 \\ 8\overline{)3.00} \\ 2\,4 \\ \hline 60 \\ 56 \\ \hline 4 \end{array} \quad\rightarrow\quad \begin{array}{r} 0.375 \\ 8\overline{)3.000} \\ 2\,4 \\ \hline 60 \\ 56 \\ \hline 40 \\ 40 \\ \hline 0 \end{array}$$

Again, we show: 3/8 = 0.375

8.2. Convert Fractions to Percents

Fractions→Percents

Example:

$$\frac{1}{100} = 1 \text{ PER}\underline{100} = 1 \text{ PER}\underline{\text{CENT}}$$

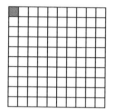

1/100 = 1 per100 = 1 perCENT = 1%

The symbol for percent is **%**

When working with % remember: **perCENT = per100 =** $\frac{1}{100}$

Here are two ways to write a fraction as a percent:

1. Create an equivalent fraction which has 100 in its
   denominator: /100
   Then because fraction /100 = perCENT, write the number
   as a percent with the % sign or the word percent.

Or:

2. Divide the numerator by the denominator. Next multiply by
   100. (You multiply by 100 since % = 1/100. See Section 8.4.)
   Then write the number with the % sign.

Example: Write 3/4 with a 100 denominator, then as a percent.

$$\frac{3}{4} = \frac{3 \times 25}{4 \times 25} = \frac{75}{100} = 75 \text{ per} 100 = 75\%$$

Example: Write 0.2/10 as a percent using equivalent fractions and using division. Then show it on a grid.

This looks odd since there is a decimal in the numerator, but we just make an equivalent fraction with 100 as the denominator. Then since /100 is per<u>cent</u>, write the number as a percent:

$$\frac{0.2 \times 10}{10 \times 10} = \frac{2}{100} = 2 \text{ per} 100 = 2\% \quad \textbf{Therefore, 0.2/10 = 2\%}$$

Alternatively you can convert by dividing 0.2 by 10:

$$10\overline{)0.02 \atop .20} \atop \underline{20} \atop 0$$

Then multiply by 100 and write %. (Multiply 100 since % = 1/100.)
0.02 x 100 = 2%. **Again, 0.2/10 = 2%**

You can see this on a grid:

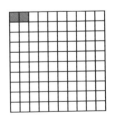

0.2/10 = 2/100

2 of the 100 boxes are filled in.

2/100 = 2 per100 = 2 perCENT = 2%

Example: Write 14/25 as a percent using equivalent fractions and division.

Write equivalent fraction with a denominator of 100. Then, because /100 is perCENT, write the number as a percent:

$$\frac{14}{25} \begin{smallmatrix} \times 4 \\ \times 4 \end{smallmatrix} = \frac{56}{100} = 56 \text{ per100} = 56 \text{ perCENT} = 56\%$$

Therefore, 14/25 = 56%

14/25 = 56/100

56 of the 100 boxes are filled in.

56/100 = 56 per100 = 56 perCENT = 56%

Alternatively, we can divide 14 by 25 to find its decimal form:

$$
\begin{array}{r}
0.5 \\
25)\overline{14.0} \\
12\ 5 \\
\hline
1\ 5
\end{array}
\quad \rightarrow \quad
\begin{array}{r}
0.56 \\
25)\overline{14.0\,0} \\
12\ 5 \\
\hline
1\ 5\ 0 \\
1\ 5\ 0 \\
\hline
0
\end{array}
$$

Then multiply by 100 since % = 1/100: 0.56 x 100 = 56%.

Again, 14/25 = 56%

Example: Write 2/5 as a percent and represent on a grid.

We can use equivalent fractions: $\dfrac{2}{5} = \dfrac{?}{100}$

Since 5 x 20 = 100 for the denominator, multiply by 20/20:

$$\frac{2}{5} = \frac{2}{5}\frac{\times 20}{\times 20} = \frac{40}{100} = 40\ per100 = 40\ perCENT = 40\%$$

Therefore, 2/5 = 40%

You can see this on a grid:

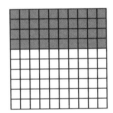

2/5 = 40/100

40 of the 100 boxes are filled in.

40/100 = 40 per100 = 40 perCENT = 40%

Again, we can divide 2 by 5 to find its decimal form:

```
    0.4
5)2.0
    2 0
    ___
       0
```

Then multiply by 100 since % = 1/100: 0.4 x 100 = 40%.
Again, 2/5 = 40%

8.3. Convert Decimals to Fractions

Decimals→Fractions

Examples:

$$0.0\underline{1} \; = \; \frac{1}{100}$$

↑

hundredths place

$$0.\underline{5} \; = \; \frac{5}{10}$$

↑

tenths place

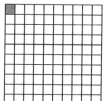

1 hundredth
1 per100
1/100

5 tenths
5 per10
5/10

A decimal is another way to write a fraction.

A fraction is another way to write a decimal.

Let's review the decimals places:

$$7,6\;5\;4,3\;2\;1\;.\;1\;2\;3\;4\;5\;6$$

| Millions | Hundred Thousands | Ten Thousands | Thousands | Hundreds | Tens | Ones | Decimal point | Tenths | Hundredths | Thousandths | Ten Thousandths | Hundred Thousandths | Millionths |

Digits on the right side of a decimal point are often called **decimal fractions**. Examples include the 2 in 0.2 and the 75 in 0.75.

Decimal fractions can be written as fractions having a denominator that is 10, 100, 1,000, 10,000, etc., such as:

$$0.75 = \frac{75}{100} = \frac{3}{4}$$

You can write a decimal as a fraction by writing
the **numerator** as the digits and
the **denominator** as the tenth, hundredth, or other decimal place occupied by the right-hand digit.

In other words, to convert a decimal into a fraction:

1. Write the digits of the decimal fraction as the numerator and place them over the 10th, 100th, 1,000th, etc., that corresponds to the far-right digit.

2. Then reduce the fraction if needed.

Example: Write and show decimal 0.5 as a fraction.

In decimal 0.5 the digit is 5 and is the furthest right digit in the 10ths place.

$$0.\underline{5} \ = \ \frac{5}{10} \ = \ \frac{5 \ \div 5}{10 \ \div 5} \ = \ \frac{1}{2}$$

0.5 equals 5 tenths, which is 5/10

We can reduce fraction 5/10 to 1/2

0.5 = 5 tenths = 5/10 = 1/2

5 of the 10 rectangles are shaded.

Example: Write decimals 2.75, 0.1234, and 1.001001 as fractions.

2.75 has digits **275**:

$$2.7\underline{5} \;=\; \frac{275}{100} \;=\; \frac{275 \div 25}{100 \div 25} \;=\; \frac{11}{4} \;\; \text{or} \;\; 2\frac{3}{4}$$

Far-right digit is in the hundredths place.

2.75 = 275 hundredths = 275/100

We can reduce fraction 275/100 to 11/4

0.1234 has digits **1234**:

$$0.123\underline{4} \;=\; \frac{1{,}234}{10{,}000}$$

Far-right digit is in the ten-thousandths place.

0.1234 = 1,234 ten-thousandths = 1,234/10,000

Note: 1,234/10,000 = 617/5,000

1.001001 has digits **1001001**:

$$1.00100\underline{1} \;=\; \frac{1{,}001{,}001}{1{,}000{,}000}$$

Far-right digit is in the millionths place.

1.001001 = 1,001,001 millionths = 1,001,001/1,000,000

8.4. Convert Decimals to Percents

Decimals→Percents

Example:

$$0.0\underline{1} \ = \ 1 \ PER\underline{100} \ = \ 1 \ PER\underline{CENT}$$

↑
hundredths place

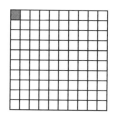

0.01 = 1 hundredth = 1 per100 = 1 perCENT = 1%

1 of the 100 squares is shaded.

Remember the symbol for percent is: **%**

Also keep in mind that: per<u>cent</u> $= \dfrac{1}{100}$

To write a decimal as a percent you simply:

1. **Multiply the decimal by 100.**

   Multiplying by 100 results in the **decimal point moving to the right 2 spaces.**

Why we can write a decimal as a percent by just multiplying by 100 (which moves the decimal point to the right 2 places)?

First, any number, even a decimal, can be written as: $\dfrac{\text{number}}{1}$

If we multiply a number by $\dfrac{100}{100}$ we don't change its value:

$$\frac{\text{number}}{1} \times \frac{100}{100} = \frac{\text{number} \times 100\!\!\!\diagup}{1 \times 100\!\!\!\diagup} = \frac{\text{number}}{1}$$

Because % sign $= \dfrac{1}{100}$, when we convert a number (or a decimal)

into a percent, we must account for "% sign $= \dfrac{1}{100}$" by also

multiplying by 100. We can see this accounting by writing:

$$\frac{\text{number}}{1} \times \frac{100}{100} = \frac{\text{number}}{1} \times \frac{100}{1} \times \frac{1}{100} = \text{number}$$

This means we can
replace the % sign with $\dfrac{1}{100}$ if we also multiply by $\dfrac{100}{1}$

In summary:

$$\frac{\text{number}}{1} \times \frac{100}{100} = \frac{\text{number}}{1} \times \frac{100}{1} \times \frac{1}{100} = \textbf{number} \times \textbf{100} \times \textbf{\%}$$

This is also true if the number is a **decimal**:

$$\frac{\text{decimal}}{1} \times \frac{100}{100} = \frac{\text{decimal}}{1} \times \frac{100}{1} \times \frac{1}{100} = \textbf{decimal} \times \textbf{100} \times \textbf{\%}$$

Example: Convert the number 3 to its pe<u>rcent</u> form.

$$\frac{3}{1} \times \frac{100}{100} = \frac{3}{1} \times \frac{100}{1} \times \frac{1}{100} = 3 \times 100 \times \% = \mathbf{300\%}$$

Example: Convert decimal 0.03 to its % form.

$$\frac{0.03}{1} \times \frac{100}{100} = \frac{0.03}{1} \times \frac{100}{1} \times \frac{1}{100} = 0.03 \times 100 \times \% = \mathbf{3\%}$$

Note that the decimal point moves 2 places to the right when 0.03 is multiplied by 100 to get 3. We see that **percents are always 100 times bigger than equivalent decimals.** The decimal point moves 2 places to the right from decimal to percent form:

$$0.03. \longrightarrow 3\%$$

x 100%

Let's look at different ways to write and draw 0.03:

$$0.0\underline{3} = \frac{3}{100} = 3 \text{ PER}\underline{100} = 3 \text{ PER}\underline{CENT} = 3\%$$

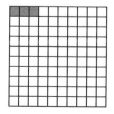

0.0<u>3</u> = 3 hundredth = 3 per100
= 3 perCENT = 3%

Example: Write decimal 0.25 as a percent.

$$0.2\underline{5} \;=\; \frac{25}{100} \;=\; 25\ \text{per100} \;=\; 25\ \text{percent} \;=\; \mathbf{25\%}$$

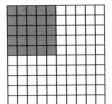

0.2<u>5</u> = 25 hundredths = 25 per100

= 25/100 = 25 perCENT = 25%

25 of the 100 squares are shaded.

We can also convert using:

$$\frac{0.25}{1} \times \frac{100}{100} \;=\; \frac{0.25}{1} \times \frac{100}{1} \times \frac{1}{100} \;=\; 0.25 \times 100 \times \% \;=\; 25\%$$

Example: Write decimal 0.2 as a percent.

$$0.2 \;=\; 0.2\underline{0} \;=\; \frac{20}{100} \;=\; 20\ \text{per100} \;=\; 20\ \text{percent} \;=\; \mathbf{20\%}$$

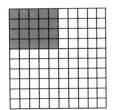

0.2<u>0</u> = 20 hundredths = 20 per100

= 20/100 = 20 perCENT = 20%

20 of the 100 squares are shaded.

We can also convert using:

$$\frac{0.2}{1} \times \frac{100}{100} \;=\; \frac{0.2}{1} \times \frac{100}{1} \times \frac{1}{100} \;=\; 0.2 \times 100 \times \% \;=\; 20\%$$

Example: Write 0.28, 5.65, and 0.0089 as percents by multiplying by 100, which moves decimal point right 2 places.

$$0.28 = \frac{0.28}{1} \times \frac{100}{100} = \frac{0.28}{1} \times \frac{100}{1} \times \frac{1}{100} = 0.28 \times 100 \times \% = \mathbf{28\%}$$

0.28. $\longrightarrow$ 28%
x 100%

Also note: 0.2<u>8</u> = 28 hundredths = 28 per100 = 28%

$$5.65 = \frac{5.65}{1} \times \frac{100}{100} = \frac{5.65}{1} \times \frac{100}{1} \times \frac{1}{100} = 5.65 \times 100 \times \% = \mathbf{565\%}$$

5.65. $\longrightarrow$ 565%
x 100%

Also note: 5.6<u>5</u> = 565 hundredths = 565 per100 = 565%

$$0.0089 = \frac{0.0089}{1} \times \frac{100}{100} = \frac{0.0089}{1} \times \frac{100}{1} \times \frac{1}{100} = 0.0089 \times 100 \times \% = \mathbf{0.89\%}$$

0.00.89 $\longrightarrow$ 0.89%
x 100%

Also note: 0.0<u>0</u>89 = 0.89 hundredths = 0.89 per100 = 0.89%

8.5. Convert Percents to Fractions

Percents→Fractions

Example:

$$1 \text{ PER}\underline{\text{CENT}} = 1 \text{ PER}\underline{100} = \frac{1}{100}$$

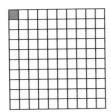

1% = 1 perCENT = 1 per100 = 1/100

1 shaded square per 100 squares.

Remember the symbol for percent is **%**

Also keep in mind that: **per$\underline{\text{cent}}$** $= \dfrac{1}{100}$

To write a percent as a fraction:

1. Just show the per$\underline{\text{cent}}$ sign as fraction /100

by writing $\dfrac{\text{number}}{100}$. Then reduce if needed.

For example, 28% $= \dfrac{28}{100}$ which reduces to $\dfrac{7}{25}$

Example: Write 20% as a fraction.

20% is 20 perCENT = 20 per100 = $\dfrac{20}{100}$ = $\dfrac{20}{100}$ = $\dfrac{2}{10}$ = $\dfrac{1}{5}$

When you see % you can think of "% = /100" by setting the number over 100. In this case:

$$20\% \;=\; \frac{20}{100} \;=\; \frac{2}{10} \;=\; \frac{1}{5}$$

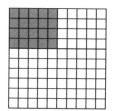

20% = 20 perCENT = 20 per100 = 20/100

20 of the 100 squares are shaded.

Example: Convert 2%, 0.323%, and 350% to fractions.

We just write /100 for %, or divide by 100!

2% = 2 perCENT = 2 per100 = $\dfrac{2}{100}$ = $\dfrac{1}{50}$

0.323% = 0.323 perCENT = 0.323 per100 = $\dfrac{0.323}{100}$ = $\dfrac{323}{100{,}000}$

350% = 350 perCENT = 350 per100 = $\dfrac{350}{100}$ = $\dfrac{35}{10}$ = $\dfrac{7}{2}$ or $3\dfrac{1}{2}$

8.6. Convert Percents to Decimals

Percents→Decimals

Example:

$$1 \text{ PER\underline{CENT}} = 1 \text{ PER}\underline{100} = 0.0\underline{1}$$

hundredths place

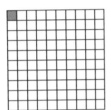

1% = 1 perCENT = 1 per100

= 1 hundredth = 0.01

1 shaded square per 100 squares.

Remember, the word per<u>cent</u> = $\dfrac{1}{100}$

And the decimals places are:

$$7,654,321 \, . \, 123456$$

Millions
Hundred Thousands
Ten Thousands
Thousands
Hundreds
Tens
Ones
Decimal point
Tenths
Hundredths
Thousandths
Ten Thousandths
Hundred Thousandths
Millionths

To write a percent as a decimal:

1. Divide the decimal by 100, since % = /100.
    This results in the decimal point moving left 2 places.

For example, 78% written as a decimal is:

$$78\% \; = \; \frac{78}{100} \; = \; 0.78$$

hundredths place

The reason you just divide the percent by 100 is:

$$\% \; = \; \text{perCENT} \; = \; /100$$

So when we write a percent as a decimal we are:

1. Writing it first as a fraction: /100

2. Then because fraction /100 makes the number a decimal hundredth (0.01), we can write it as a decimal hundredth.

Example: Write 20% in its decimal form.

First, we know 20% is 20 per100 which we can write: $\dfrac{20}{100}$

Because fraction /100 makes the number a decimal hundredth (0.01), we can write 20/100 it as decimal hundredth 0.20
So:

$$20\% \ = \ \frac{20}{100} \ = \ 0.2\underline{0}$$

Therefore, 20% in its decimal form is: 0.20

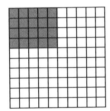

20% = 20 perCENT = 20 per100

= 20/100 = 20 hundredths = 0.20

20 of the 100 squares are shaded.

Note: We can divide 20 by 100 to prove $\dfrac{20}{100} \ = \ 0.20$:

```
        .2
100)20.0
     20 0
        0
```

Example: Write 3% in its decimal form.

First, we know 3% is 3 per100 which we can write: $\dfrac{3}{100}$

Because fraction /100 makes the number a decimal hundredth (0.01), we can write 3/100 it as decimal hundredth 0.03
So:

$$3\% \;=\; \dfrac{3}{100} \;=\; 0.0\underline{3}$$

Therefore, 3% in its decimal form is: 0.03

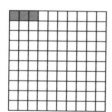

3% = 3 perCENT = 3 per100

= 3 hundredths = 0.03

3 shaded squares per 100 squares

Note: We can divide 3 by 100 to prove $\dfrac{3}{100}$ = 0.03:

$$\begin{array}{r} 0.03 \\ 100{\overline{)3.00}} \\ \underline{3\ 00} \\ 0 \end{array}$$

Example: Write 75%, 7.5%, 0.75%, and 750% as decimals.

First, we know % = per100 = /100
Because /100 makes the number a decimal hundredth (0.01),
we can write it as a decimal hundredth.
Remember, dividing by 100 moves decimal point left 2 places.

75% = 75 per100 = $\dfrac{75}{100}$ = .75. = **0.75**

/100

7.5% = 7.5 per100 = $\dfrac{7.5}{100}$ = 0.07.5 = **0.075**

/100

0.75% = 0.75 per100 = $\dfrac{0.75}{100}$ = 0.00.75 = **0.0075**

/100

750% = 750 per100 = $\dfrac{750}{100}$ = 7.50. = **7.50**

/100

8.7. Practice Problems

8.1

(a) Students living more than 0.8 miles from school may take the bus. Bill remembers he lives 6 blocks from school and each block is 1/8 mile long. Can he take the bus?

(b) Express the following as decimals: $\dfrac{10}{25}$, $\dfrac{21}{25}$, $\dfrac{42}{25}$

8.2

(a) To get an A you must get at least 90% correct answers on a test. You missed 3 of 40 questions. Did you get an A?

(b) Express the following as percents: $3\dfrac{3}{4}$, $\dfrac{1}{100,000}$, $\dfrac{1}{6}$

8.3

(a) Using a micrometer, you measure the diameter of a steel ball as 0.3125 inches. What is this expressed as a fraction?

(b) Express the following as fractions: 0.002, 0.125, 2.075

8.4 Convert the following to percents: 0.0072, 1.23, 0.913

8.5

(a) You received 62.5% of the vote. What fraction did you get?

(b) Convert these percents to fractions: 175%, 0.5%, 1.25%

8.6

(a) Your one-meter long measuring stick shows intervals down to millimeters (thousandths of a meter). Your friend cuts off 36% of the stick. How many millimeters long is the stick now?

(b) Write these percents as decimals: 999%, 32.10%, 5.2%

Answers to Chapter 8 Practice Problems

8.1

(a) Bill lives 6 x 1/8 = 6/8 or 3/4 miles from school.

Using equivalent factions: $\dfrac{3}{4} \times \dfrac{25}{25} = \dfrac{75}{100} = 0.75$ miles.

Students must live more than 0.8 miles from school to take the bus. Bill lives 0.75 miles away so he **cannot take the bus**.

We can check our answer of 0.75 miles using division. Just divide 3/4 to find the decimal form:

$$
\begin{array}{r}
0.7 \\
4\overline{)3.00} \\
2\,8 \\
\hline
20
\end{array}
\qquad \longrightarrow \qquad
\begin{array}{r}
0.75 \\
4\overline{)3.00} \\
2\,8 \\
\hline
20 \\
20 \\
\hline
0
\end{array}
$$

Again, 3/4 = 0.75 miles, so Bill **cannot take the bus**.

(b)

$$\frac{10}{25} \times \frac{4}{4} = \frac{40}{100} = \mathbf{0.40}$$

$$\frac{21}{25} \times \frac{4}{4} = \frac{84}{100} = \mathbf{0.84}$$

$$\frac{42}{25} \times \frac{4}{4} = \frac{168}{100} = \mathbf{1.68}$$

8.2

(a) 40 – 3 = 37 correct answers.

You can find % using: $\dfrac{37}{40} = \dfrac{?}{100}$:

$\dfrac{37}{40} \times \dfrac{25}{25} = \dfrac{925}{1,000} = \dfrac{92.5}{100} = 92.5\%$. **Yes**. Congratulations on your A!

(b) Remember: /100 = %

$3\dfrac{3}{4} = \dfrac{15}{4} = \dfrac{15}{4} \times \dfrac{25}{25} = \dfrac{375}{100} =$ **375%**

$\dfrac{1}{100,000} = \dfrac{1}{100,000} \dfrac{\div 1,000}{\div 1,000} = \dfrac{\frac{1}{1,000}}{100} = \dfrac{1}{1,000}\% =$ **0.001%**

thousandths place

$\dfrac{1}{6}$ can be converted to a decimal using long division:

$$\begin{array}{r} 0.1 \\ 6\overline{)1.00} \\ \underline{0\ 6} \\ 40 \end{array} \longrightarrow \begin{array}{r} 0.166 \\ 6\overline{)1.000} \\ \underline{0\ 6} \\ 40 \\ \underline{36} \\ 40 \\ \underline{36} \end{array} \longrightarrow$$

As we continue the long division we keep getting 6's in the answer...

So $\dfrac{1}{6}$ is approximately 0.1667

To write the decimal as a percent remember: $\% = \dfrac{1}{100}$

So we must also multiply the numerator by 100.
Therefore, 0.1667 x 100 = **16.67%**

8.3

(a) We write the digits of the decimal over the 10th, 100th, etc., place that corresponds to the far-right digit. For 0.3125 inches, the far right digit is in the 10,000s place. So:

$$0.3125 = \frac{3{,}125}{10{,}000} \quad \text{We can reduce this: } \quad \frac{3{,}125}{10{,}000} \frac{\div 625}{\div 625} = \frac{5}{16} \text{ inches}$$

(b) Note that each right digit is in the thousandths place.

$$0.002 = \frac{2}{1{,}000} = \frac{1}{500}, \quad 0.125 = \frac{125}{1{,}000} = \frac{1}{8}, \quad 2.075 = \frac{2{,}075}{1{,}000} = \frac{83}{40} \text{ or } 2\frac{3}{40}$$

8.4

0.0072 = 0.72%, 1.23 = 123%, 0.913 = 91.3%. Let's see why:

$$0.0072 = \frac{0.0072}{1} \times \frac{100}{100} = 0.0072 \times 100 \times \% = \mathbf{0.72\%}$$

    Note that multiplying by 100 moves decimal point right 2 places.

$$1.23 = \frac{1.23}{1} \times \frac{100}{100} = 1.23 \times 100 \times \% = \mathbf{123\%}$$

    Note that multiplying by 100 moves decimal point right 2 places.

  Also notice: 1.23 = 123 hundredths = 123 per100 = 123%

$$0.913 = \frac{0.913}{1} \times \frac{100}{100} = 0.913 \times 100 \times \% = \mathbf{91.3\%}$$

    Note that multiplying by 100 moves decimal point right 2 places.

8.5

(a) 62.5% = 62.5 per100 = $\dfrac{62.5}{100} = \dfrac{625}{1,000} = \dfrac{625 \div 125}{1,000 \div 125} = \dfrac{5}{8}$ of vote

(b) 175% = 175 per100 = $\dfrac{175}{100} = \dfrac{175 \div 25}{100 \div 25} = \dfrac{7}{4} = 1\dfrac{3}{4}$

0.5% = 0.5 per100 = $\dfrac{0.5}{100} = \dfrac{0.5 \times 2}{100 \times 2} = \dfrac{1}{\mathbf{200}}$

1.25% = 1.25 per100 = $\dfrac{1.25}{100} = \dfrac{1.25 \times 4}{100 \times 4} = \dfrac{5}{400} = \dfrac{5 \div 5}{400 \div 5} = \dfrac{1}{\mathbf{80}}$

8.6

(a) 1 meter (m) is 100% of stick length.

100% - 36% = 64%. Because % = per100:

64% = 64 per100 = $\dfrac{64}{100}$ = .64. = 0.64 meters

/100

A millimeter is $\dfrac{1}{1,000}$ of a meter, so 0.640 m = **640 millimeters**

(b)

$999\% = 999 \text{ per} 100 = \dfrac{999}{100} = 9.99. = \mathbf{9.99}$ ↖ hundredths place

/100

$32.10\% = 32.10 \text{ per} 100 = \dfrac{32.10}{100} = .32.10 = \mathbf{0.3210}$

/100

$5.2\% = 5.2 \text{ per} 100 = \dfrac{5.2}{100} = .05.2 = \mathbf{0.052}$

/100

Extra Credit Section: The Base 2 System

In all thy ways acknowledge him, and he shall direct thy paths. Proverbs 3:6

Have you wondered why we use a **Base 10 decimal** number system? Remember, in the Base-10 decimal system the value of each place goes up by 10 times as you move from **right to left** (ones, tens, hundreds, etc.). Perhaps the Base 10 system developed that way because people have 10 fingers. Of course, you could invent a number system using other bases, like 5 or 16.

The Base 2 Binary System Used by Computers

What about computers? What number or counting system do they use? In computers the smallest unit of its memory, a "bit", is a magnetic spot that is either "on" or "off", + or -, 1 or 0. Because computer bits use only 1s and 0s, they are designed to work with a **Base 2 binary number system**. In a Base 2 system the only two possible values are **0** and **1**, and everything is based on the power (or exponent) of 2. This means the value of each place increases by 2 times (or a power of 2) as you move from **right to left**. Here are the first eight Base 2 binary place values. See how each one is 2 times greater:

2^0 = **1** any number to the power of 0 is 1
2^1 = **2** any number to the power of 1 is itself
2^2 = **4** = 2 x 2
2^3 = **8** = 2 x 2 x 2
2^4 = **16** = 2 x 2 x 2 x 2
2^5 = **32** = 2 x 2 x 2 x 2 x 2
2^6 = **64** = 2 x 2 x 2 x 2 x 2 x 2
2^7 = **128** = 2 x 2 x 2 x 2 x 2 x 2 x 2

Remember, in the Base 10 system, where there are ten place values, you can count from 0 to 9 in each place. In the Base 2 binary system, where there are only two place values, 0 and 1, you can count from 0 to 1 in each place. Therefore, Base 2 numbers are made up of 0s and 1s. Let's look at a Base 2 number with its first 8 places as 1s. Note how its columns are valued:

$$1 \quad 1 \quad 1 \quad 1 \quad 1 \quad 1 \quad 1 \quad 1.$$

| One Hundred Twenty-eights | Sixty-fours | Thirty-twos | Sixteens | Eights | Fours | Twos | Ones |

$$128 + 64 + 32 + 16 + 8 + 4 + 2 + 1 = 255$$

$$2^7 + 2^6 + 2^5 + 2^4 + 2^3 + 2^2 + 2^1 + 2^0 = 255$$

You can see that 11111111 in binary Base 2 is 255 in Base 10.

We can begin to see how binary Base 2 numbers relate to decimal Base 10 numbers. Let's keep it simple and just write the first **five** places using columns beginning with 1:

2^4	2^3	2^2	2^1	2^0	=	$16 + 8 + 4 + 2 + 1$	=	31
16	8	4	2	1	=	$16 + 8 + 4 + 2 + 1$	=	31
1	1	1	1	1	=	Binary = 11111, Note: each "1" shows a value other than 0 exists.		

See that Base 2 number 11111 is 31 in Base 10.

If you're still a little confused keep reading--you'll SEE it!

Let's convert Base 2 binary number 10101 to Base 10.

Just write the binary number in columns and fill in zero where there is a **0** and the Base 10 place value where there is a **1**:

Binary: **1**   **0**   **1**   **0**   **1**

2^4 2^3 2^2 2^1 2^0

Decimal: 16 + 0 + 4 + 0 + 1 = 21

We see that binary number 10101 is 21 in Base 10.

Now convert Base 2 binary number 11010110 to Base 10.

Again write the binary number in columns and fill in zero where there is a **0** and the Base 10 place value where there is a **1**:

Binary: **1**   **1**   **0**   **1**   **0**   **1**   **1**   **0**

2^7 2^6 2^5 2^4 2^3 2^2 2^1 2^0

Decimal: 128 + 64 + 0 + 16 + 0 + 4 + 2 + 0 = 214

We see that binary number 11010110 is 214 in Base 10.

How do you convert from Base 10 decimal to Base 2 binary?

Let's look at the number 3:   2 + 1 = 3

You can also write this as: $2^1 + 2^0 = 3$

To translate into binary put **1**'s or **0**'s in the columns under the Base 2 numbers. In this case we have values (a 2 and a 1) in the 2^1 and 2^0 columns, so put **1s** in the 2^1 and the 2^0 columns:

$2^1 + 2^0 = 3$

1   **1**

We see that the decimal number 3 written in binary is: 11

Let's work through a general process on how to convert from Base 10 to Base 2 by converting decimal number 20 to binary:

First look for the largest Base 2 place value that is **less than or equal to 20**. See that 20 is between 2^4 = 16 and 2^5 = 32.

So the largest power-of-2 less than or equal to 20 is: 2^4 = **16**

Write the Base 2 columns beginning with 2^4 and put a **1** under 2^4:

16	8	4	2	1
2^4	2^3	2^2	2^1	2^0
1				

What are we left with? To find out subtract 2^4 = 16 from 20: 20 - 16 = 4. We are left with 4.

Now look for the largest Base 2 place value **less than or equal to 4**. We see that 2^2 = 4.

So the largest power-of-2 less than or equal to 4 is: 2^2 = **4**

Put **1** in the 2^2 column. Also fill in a **0** in the 2^3 column:

16	8	4	2	1
2^4	2^3	2^2	2^1	2^0
1	0	1		

What are we left with? Subtract 4 from 4: 4 - 4 = 0

Is there a Base 2 place value **less than or equal to 0**?

No. The smallest is 2^0 = 1. There are no values left.

Fill in **0**'s for the 2^1 and 2^0 columns:

16	8	4	2	1
2^4	2^3	2^2	2^1	2^0
1	0	1	0	0

So decimal number 20 converts to binary number: 10100

Thank you for working though *I SEE MATH: THE BASICS*!

We hope you enjoyed it and learned from its unique, visual approach to SEEING math.

The next *I SEE MATH* book for those who want to go further will be/is called:

I SEE MATH: BUILDING ON THE BASICS

Keep up the good work and remember,

YOU REALLY HAVE MATH BRILLIANCE!

"Mathematics is the language with which God has written the universe."
Galileo Galilei

About the Authors

Debra Anne Ross Lawrence earned a double BA in Biology and Chemistry with honors from the University of California at Santa Cruz and an MS in Chemical Engineering from Stanford University. Her career includes R&D management in engineering, biosensors, pharmaceutical drug discovery, and intellectual property.

David Allen Lawrence earned BS and MS degrees in Management from the Massachusetts Institute of Technology and a Juris Doctor *cum laude* from the University of Minnesota. His career includes Director of Law for a Fortune 500 energy company, Senior Vice President and General Counsel for an engineering and consulting company, and Chief Judge for a state regulatory commission.

David and **Debra** currently reside in Alaska. Their extra-literary activities include all-season mountaineering, studying Bible prophecy, pursuing an ultra-healthy lifestyle, organic gardening, following their dreams, and living together happily forever. David and Debra enjoy receiving questions/comments: glacierdog@glacierdogpublishing.com

David and Debra's Books include:

I See Math: The Basics
Skippy & Sam Unleashed: The Year Everything Changed
Arrows Through Time: A Time Travel Tale of Adventure, Courage, and Faith
The 3:00 PM Secret: Live Slim and Strong Live Your Dreams
The 3:00 PM Secret 10-Day Dream Diet
Master Math books: *Basic Math and Pre-Algebra, Algebra, Geometry, Pre-Calculus, Trigonometry, Calculus,* and *Essential Physics*

Jesus said unto him,
Thou shalt love the Lord thy God
with all thy heart, and with all thy soul,
and with all thy mind. This is the first and
great commandment.
Matthew 22:37-38

Index

*But even the very hairs of your head are all **numbered**. Fear not... Luke 12:7*

Books by Debra and David Lawrence

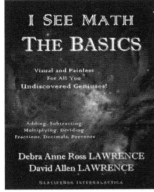

GlacierDog

Publishing

glacierdogpublishing.com

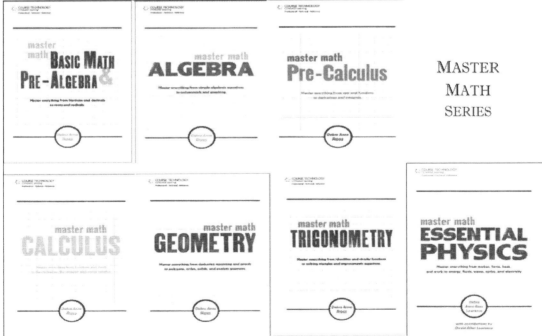

Master

Math

Series

For I am persuaded, that neither death, nor life, nor angels, nor principalities, nor powers, nor things present, nor things to come, Nor height, nor depth, nor any other creature, shall be able to separate us from the love of God, which is in Christ Jesus our Lord. Rom 8:38-9

70948251R00208

Made in the USA
San Bernardino, CA
09 March 2018